Chemistry and Modern Society

Chemistry and Modern Society

Historical Essays in Honor of Aaron J. Ihde

John Parascandola, EDITOR
University of Wisconsin

James C. Whorton, EDITOR
University of Washington

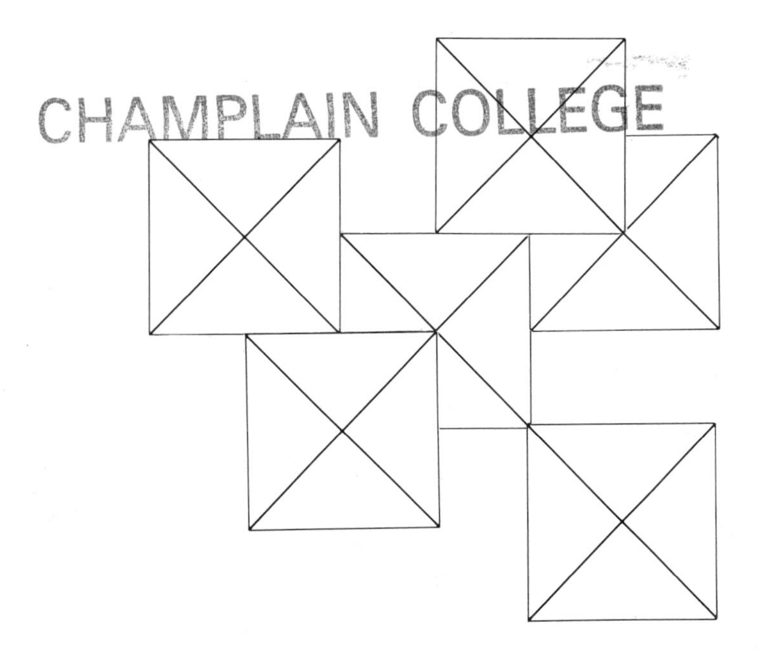

ACS SYMPOSIUM SERIES 228

AMERICAN CHEMICAL SOCIETY
WASHINGTON, D.C. 1983

Library of Congress Cataloging in Publication Data

Chemistry and modern society.

(ACS symposium series, ISSN 0097-6156;228)

Bibliography: p.
Includes index.

1. Chemistry—History—Congresses. 2. Ihde, Aaron John, 1909- . I. Ihde, Aaron John, 1909- . II. Parascandola, John. III. Whorton, James, 1942- IV. American Chemical Society. V. Series.

QD15.C46 1983 306′.45 83-11740
ISBN 0-8412-0795-X
ISBN 0-8412-0803-4 (pbk)

PRINTED IN THE UNITED STATES OF AMERICA

ACS Symposium Series

M. Joan Comstock, *Series Editor*

FOREWORD

The ACS SYMPOSIUM SERIES was founded in 1974 to provide a medium for publishing symposia quickly in book form. The format of the Series parallels that of the continuing ADVANCES IN CHEMISTRY SERIES except that in order to save time the papers are not typeset but are reproduced as they are submitted by the authors in camera-ready form. Papers are reviewed under the supervision of the Editors with the assistance of the Series Advisory Board and are selected to maintain the integrity of the symposia; however, verbatim reproductions of previously published papers are not accepted. Both reviews and reports of research are acceptable since symposia may embrace both types of presentation.

CONTENTS

Professor Ihde in his office at the University of Wisconsin about 1964

PREFACE

AARON IHDE'S CONTRIBUTIONS to our understanding of the evolution of chemistry are remarkable nearly as much for their breadth as for their quantity and quality.

This volume originated in a desire to honor the dean of American historians of chemistry, Aaron J. Ihde, at the time of his retirement from the University of Wisconsin faculty in 1980. From the beginning, however, the editors were determined to avoid the typical "Festschrift" volume consisting of a collection of papers on miscellaneous subjects held together only by the association of the authors as students or colleagues of the honoree. Convinced that a book of essays with a central unifying theme would be a more valuable contribution to the literature we strove to achieve that objective. More as a result of chance than editorial wisdom, we have succeeded to a greater extent than we originally dared hope. The result is this publication on "Chemistry and Modern Society".

A bibliography of Idhe's contibutions (*see* appendix) which ranges from Boyle's definition of the element to the establishment of food standards, from Paracelsus to pesticides, reflects something more than Idhe's extra-ordinary curiosity. It expresses a rich appreciation of the inseparability of theoretical and technical progress in chemistry, and of the participation of each in intellectual and social history. That last topic, the impact of chemistry on its social environment, has long been a special interest of Professor Ihde's, and he and his students have made significant contributions to the literature on this subject. Indeed, his chief work, "The Development of Modern Chemistry," opens with the announcement that one of its purposes is to provide "an object lesson of the role of pure science in the development of technology, agriculture, and medicine." Subsequent chapters fulfill that promise, exploring chemistry's influence on western society's transformation from an agricultural to a high-technology industrial base with more thorough and up-to-date coverage than any other broad survey of chemical history. The purpose of the present volume is to modestly offer in tribute a few additional object lessons illuminating chemistry's diverse roles in modern civilization, focusing especially on the American scene.

The theme of Professor Ihde's Dexter Award address, the formation of hybrid sciences, is extended by an analysis of the conceptual origins of geochemistry. This analysis reveals how the maturation of physical chemistry in the late nineteenth century provided access to the internal dynamics of geological processes and spawned a new specialty in which American scientists took the lead. A more pragmatic application of chemistry to geological deposits (and one having renewed relevance of late) is examined in a detailed study of the history of the production of synthetic petroleum from coal. Petroleum synthesis, of course, has been largely an exercise in chemical engineering, another hybrid discipline. A crucial stage in that profession's conceptual development is dealt with in a discussion of the early twentieth century utilization of the notions of unit operations and unit processes.

Another area in which chemistry has had an important impact on modern society is in the field of health. Professor Ihde has been a major contributor to the history of nutritional biochemistry, and hence it is appropriate that this volume includes a paper dealing with the early history of vitamin research. A somewhat different merger between chemistry and the health sciences has been the pursuit of pharmaceutical investigation, an activity that, however thriving now, at first took frequently halting steps. The checkered fortunes of Charles Holmes Herty are striking testimony to the difficulties of organizing sophisticated drug research in the 1920s.

The mention of drug research is a reminder that the career of chemistry has been checkered during the twentieth century. The term "drug therapy" is nearly as likely to trigger the response "thalidomide" as to make one think of penicillin. As the side effects of chemical progress have become more apparent, the whole science has come to be regarded as the great benefactor as well as the great despoiler, a symbol of modern society's paradox of unprecedented prosperity shadowed by the potential for unparalleled devastation. "The Development of Modern Chemistry" concludes on just such a note, with the ominous injunction that although chemistry "can still do much for mankind," it "can also be his undoing." The double-edged action of chemistry on society is the minor theme running through this volume. The 1937 Elixir Sulfanilamide tragedy, for example, serves as a dramatic backdrop for a presentation of the vicissitudes affecting drug regulation, while the establishment of standards for synthetic food colors demonstrates the uncertainties confounding the policing of substances of significant economic value but indeterminate toxicity. Diet has been contaminated by chemical theory as well. Food fads such as the uric acid fetish have been legitimized in practitioners' minds by self-serving distortions of biochemical principles. Finally, chemistry has altered the face of war in the twentieth century, as exemplified by the production of toxic gases for use in World War I.

The interface between chemistry and society refuses to be drawn in such neat lines as good and bad, however. The search for chemical warfare agents popularized the method of project research that would be used so fruitfully in peacetime investigations. The ravages of Elixir Sulfanilamide provoked the present stricter food and drug legislation. These case studies suggest some of the complexity, as well as the diversity, of chemistry's influence on modern civilization.

JOHN PARASCANDOLA

University of Wisconsin
Madison, Wisconsin

JAMES WHORTON

University of Washington
Seattle, Washington

March 18, 1983

ACKNOWLEDGMENTS

This book would not have been possible without the assistance of various institutions and individuals. The editors wish to express their appreciation to the following for providing the financial support necessary for the publication of this volume: Division of History of Chemistry, American Chemical Society; Merck and Company; Smith Kline Corporation; Dexter Chemical Corporation; and Warner-Lambert Company. Thanks are also due to more people than we can acknowledge here, but we would at least like to single out the following individuals for their special efforts on behalf of this project: Natalie Foster, Secretary-Treasurer, ACS Division of the History of Chemistry; and Suzanne Roethel and Antoinette Drexler, ACS Books Department.

INTRODUCTION

Aaron J. Ihde—Perspectives

AARON JOHN IHDE WAS BORN on a farm near Neenah, Wisconsin on December 31, 1909, and has lived in the badger state essentially all of his life. After receiving his B.S. in chemistry from the University of Wisconsin in 1931, Ihde spent the next seven years as a research chemist at a creamery in Chicago. During this period, he was married to Olive Jane Tipler, a former high school classmate and a teacher of Latin and history. The Ihdes were later to have two children, Gretchen (Serrie), now manager of and violinist in the Florida West Coast Symphony in Sarasota, and John, now a high school chemistry teacher and basketball coach in Wausau, Wisconsin.

A desire for further education led Ihde to return to Wisconsin for graduate study in 1938. He earned his M.S. in 1939 and his Ph.D. in 1941 for work in food chemistry under Professor H. A. Schuette. After a year on the faculty at Butler University, he returned to Madison once again in 1942, this time to stay, as a member of the Wisconsin chemistry faculty.

For some years Ihde's principal teaching responsibilities were in freshman chemistry, but the historical interests that were eventually to become predominant in his career soon began to surface. In 1946 he revived a History of Chemistry course that had not been taught since the retirement of Professor Lewis Kahlenberg in 1940. Two years later, Ihde's interdisciplinary leanings were given a further stimulus when he agreed to teach the Physical Universe course in the new Integrated Liberal Studies (ILS) program. The purpose of this two-year program in the College of Letters and Science was to provide a coordinated sequence of courses in humanities, social studies, and natural sciences that dealt with the development of ideas in Western culture. Ihde's course dealt with the physical sciences, and he utilized a case study approach that emphasized the historical development of key concepts in astronomy, physics, and chemistry.

A year (1951–52) spent at Harvard University as a Carnegie Fellow had a significant influence on the direction of Ihde's career. His association with James Conant, Leonard Nash, and Thomas Kuhn in the teaching of Natural Science 4 at Harvard was a valuable experience that influenced the development of his own Physical Universe course. More importantly from our present point of view, his contact with the History of Science program at Harvard, manned by George Sarton and I. B. Cohen, moved him further in the direction of history of science. Ihde's conversion from chemist to historian was also accelerated by his being given a joint appointment in the Department of History of Science at Wisconsin in 1957.

Ihde had published a number of historical papers in the late 1950s and early 1960s, but his "The Development of Modern Chemistry" (1964) firmly established his credentials as an historian of chemistry. This 850-page volume remains the standard reference work on the history of modern chemistry and is an indispensable tool for students, teachers, and researchers. This book alone assures him a lasting place in the field of history of chemistry.

A year later, in 1965, Ihde edited in conjunction with William F. Kieffer "Selected Readings in the History of Chemistry," a collection of historical papers reprinted from the *Journal of Chemical Education* and published by the Division of Chemical Education of the American Chemical Society. During this same period, he also served as Chairman of the ACS Division of History of Chemistry for three years (1962–1964). All of these contributions, but especially " The Development of Modern Chemistry," led to his receiving the Dexter Award of the Division of History of Chemistry in 1968.

Meanwhile Ihde was busy making Wisconsin into a center for research and graduate study in the history of chemistry. The first Ph.D. for historical research under Ihde's direcion was awarded to Robert Siegfried in 1952, a joint degree in History of Science and Chemistry. (Siegfried was to join the History of Science Department faculty at Wisconsin in 1963, further strengthening the history of chemistry area.) In the ensuing three decades, a total of twenty-two doctoral dissertations in the field of history of science have been directed by Aaron Ihde at the University of Wisconsin. The editors and several other contributors to this volume are proud to be included in that group.

Ihde's undergraduate teaching interests were also expanding, as he developed new courses in Science, Technology and Society (in conjunction with Victor Hilts of the History of Science Department) and Evolution of Food and Drug Controls (in conjunction with John Parascandola and Glenn Sonnedecker of the History of Pharmacy program). This latter course reflects Ihde's long-time interest in food chemistry, which began in his graduate school days. From 1955 through 1968 he was a member of the Wisconsin Food Standards Advisory Committee, serving as chairman in the period 1964–1966. Both courses illustrate his concern with the application of scientific knowledge and its social consequences, the central theme of the present volume. Ihde's Physical Universe course also came in time to focus heavily on the impact of science on the quality of life. In 1978 the University of Wisconsin recognized Professor Ihde's contributions to education by conferring upon him a distinguished teaching award.

Since his retirement in 1980, Aaron Ihde has continued to remain active as a scholar and as a teacher. He continues to participate in various courses at Wisconsin as a guest lecturer and to deliver lectures and papers at various institutions and meetings around the country. He remained as major professor to several graduate students who had not yet completed their doctoral dissertations at the time of his retirement. A book-length history of the Chemistry Department at Wisconsin, on which he has been working for the past few years, is now nearing completion.

Aaron and Olive Ihde continue to live in Madison in their charming house in the University Arboretum. Those who have visited their home have fond memories of the beautiful setting and warm hospitality that they encountered there. In recent years a number of graduate students have had the good fortune to participate in History of Chemistry evening seminars held in the Ihde home, sessions in which Olive Ihde joined Aaron and his students as an active participant. Those of us involved with the preparation of this book join with all of their other friends and associates in wishing them both many more happy and productive years.

John Parascandola
James C. Whorton
March 28, 1983

1

The Intellectual Basis of Specialization
Geochemistry in America, 1890–1915

JOHN W. SERVOS
Princeton University, Program in History of Science, Princeton, NJ 08544

Geochemistry was one of several new hybrid specialties to flourish in the United States in the twentieth century. It had its origins when geologists began to apply the tools of physical chemistry to the problems of petrogenesis and metamorphism. This paper treats the reasons why America proved a congenial site for the development of this work. Especially important was the presence in the United States of vigorous communities of petrographers and physical chemists who perceived a fit between the problems of one field and the techniques of the other and who were able to mold institutions within which the specialty might grow. The first of these institutions were established within the U.S. Geological Survey and the Carnegie Institution of Washington. Only well after 1900, when geochemists discovered unanticipated industrial applications of their knowledge, did the field find a secure place in American universities.

The years around the turn of the century are replete with examples of new scientific specialties emerging from single parent fields or, more often, from the junction of two or more established scientific disciplines. A list of these new specialties might include physical chemistry, astrophysics, biochemistry, and genetics; and the roster could be easily extended. Although most had their origins in Europe, they found especially congenial climates for expansion in the United States. Indeed, it was in these new fields rather than in older specialties, such as organic chemistry, that American scientists first excelled. The standing in their specialties of a G. N. Lewis or Thomas Hunt Morgan well illustrates this point. They were not just influential

0097–6156/83/0228–0001 $06.00/0

teachers or entrepreneurs, they were also scientists who developed new techniques, ideas, and lines of research which shaped their disciplines in important ways.

An American propensity for achievement in new, interdisciplinary fields may be viewed as part of the larger process of specialization that was such a prominent feature of American intellectual life in the late nineteenth century. Specialization, of course, is not a phenomenon that has passed without commentary. Indeed, much of what has been written during the past ten years about the history of science in America deals with it. Some historians have charted the histories of individual disciplines; others have treated the sites of specialized inquiry or the development of markets for specialists' services; still others have examined the question of how specialization could have proceeded so rapidly in a society supposedly wed to egalitarian and democratic principles (1-5). A tendency to identify the American university as the crucial institution in this process is a common denominator in this literature. The size, diversity, and flexibility of American colleges and universities, we are told, made rapid specialization possible -- indeed, fostered it. Although industries and government exerted a demand for men with esoteric forms of knowledge, it was the university that spawned and nurtured them. The elective system, the graduate school, the academic department all contributed to making the American university, in the words of John Higham, the "matrix of specialization" (6-9).

I do not wish to challenge the idea that the American university of the turn of the century was a flexible institution. Nor do I wish to reject the claim that it served as the major site of specialization. Nevertheless, recognition of the importance of the university as a locus of this phenomenon should not cause us to neglect other kinds of institutions in which new specialties coalesced. Nor should the current emphasis upon institutional forms and markets for specialized knowledge blind us to the fact that ideas too played a role -- that scientists with ideas could and did develop new specialties despite the absence of institutions or social and economic needs well suited to the realization of their disciplinary aims.

The development of geochemistry in America illustrates these points. Its history bears some similarity to the histories of other hybrid specialties that emerged around the turn of the century, but in several important ways it constitutes an exception to the normal pattern of development. It did not take shape in the university. Rather it found its first expression in the United States in the laboratories of the U.S. Geological Survey and the Carnegie Institution of Washington. Only very gradually did it find a place in the university curriculum. Nor did it grow in response to market demands. Rather it arose through the efforts of a group of chemists and geologists to answer questions that were of great intellectual interest but of no immediate economic or social significance. Only gradually did experts in the new field develop or discover applications for their special knowledge and techniques.

Despite the impracticality of the subject and the indifference of university officials toward it, American scientists made fundamental contributions to geochemistry in the early twentieth century. This paper seeks to determine why. The answer, I maintain, is not to be found in institutional arrangements or in market forces, but rather in the multiplicity of intellectual traditions represented in the United States, in the openness of the American scientific community to cross-disciplinary intellectual exchange, and in the willingness of American scientists to undertake experiments in multi-disciplinary collaboration.

Before proceeding, it is essential to discuss the history of the term 'geochemistry', for its meaning has changed over time. It, or its German equivalent, appears to have been coined in 1838 by the German natural philosopher, C. F. Schönbein, to describe the study of the chemical and physical properties of geological formations. During the subsequent forty years the word appeared sporadically in the geological and chemical literature. Often used interchangeably with the term 'chemical geology', it was generally applied to a miscellany of topics in the areas of rock and mineral analysis (10, 11, 12).

If Schönbein or his successors harbored ambitions for this field, they were disappointed, for the years around mid-century were a low point in the relations between geology and chemistry. The interest in the chemical composition of igneous rocks that had been stirred by the work of Werner, Playfair, and Hall had quieted, and geologists generally were preoccupied with maps, the stratigraphical column, and the history of land forms. Sedimentary, rather than igneous, rocks were at the focus of their inquiry; paleontological evidence seemed more valuable than chemical data. Although it was occasionally suggested that the chemical composition of igneous rocks might shed light on historical questions, the complexity of eruptive deposits resisted any simple analysis. Chemistry was a valuable tool for those interested in economic geology, and, where mining was important, the geological chemist had ample work. But chemical analysis was often quite separable from geological field work. Geologists might collect rock specimens, note their abundance, location, and field relations, but often sent their samples to trained chemists for study. In England, students could obtain degrees in geology without ever having studied chemistry, and although there were closer links between chemistry and geology in Germany and Scandinavia, the chemical literacy of geologists was usually not very high (12, 13).

The relation between chemistry and petrography, the science of the classification of rocks, was somewhat closer since chemical composition was thought by some petrographers to offer a basis for rational classification. But early efforts to categorize rocks by their chemical composition were frustrated as it was discovered that it was impossible to find simple chemical formulas for these bodies. Rocks of different mineral composition fell into the same chemical categories, and rocks that were chemically alike were sometimes quite different in structure, texture, and field relations. Moreover, the success of H. Clifton Sorby's technique for the microscopic study of thin sections of rocks accentuated the tendency to identify rocks on the basis of external rather than internal properties (14, 15).

Toward the end of the century, however, chemistry, which had been the assayer's art and the geologist's somewhat neglected handmaiden, assumed a new importance for geologists and petrographers. It ceased to be simply a reservoir of techniques the geologist called upon when in need of information about rock and mineral composition, and became instead the basis for answering fundamental questions pertaining to the origins of geological structures and the reasons for geological change, questions involving the causes and mechanisms of igneous petrogenesis, metamorphism, and sedimentation. Chemistry acquired a new stature _not_ because geologists began to ask new questions; each of these problems had a hoary history. Rather it was because of the greatly enlarged ability of chemists to handle complex geological data. In particular, the ideas and techniques of the new discipline of physical chemistry opened the way to the treatment of intricate dynamical questions that were beyond the capacity of inorganic or analytical chemists of an earlier generation. The junction of geological problems and the techniques and ideas of physical chemistry was described by a variety of names at the turn of the century: physico-chemical geology, chemical petrology, geochemistry, geophysics. What is clear, however, is that during the 1890s a loose network of chemists and geologists took form around a set of old problems and a set of new or improved tools. Out of this network there arose the first recognizably modern works on geochemistry (11, 16-19).

With the benefit of hindsight, it seems obvious that geology should have become a field of application for physical chemistry. As early as 1851, Robert Bunsen had observed that magmas were solutions and might be treated by the same principles that governed the behavior of aqueous solutions. Minerals did not simply crystallize out of magmas in the reverse order of their fusibilities, but might show the same complex solubility patterns as were exhibited by mixtures of salts in water (20, 21).

In 1857, the French petrographer, J. Durocher, drew another analogy when he noted similarities between magmas and alloys (22). If one could understand the principles governing the solidification of metallic melts, which could yield alloys with different physical properties depending on the rate and conditions of cooling, might not that understanding be extended to rock melts? Both Bunsen and Durocher were seeking to comprehend the inaccessible and complex in terms of the familiar and simpler. Their suggestions, however, bore no direct fruit, for although aqueous solutions and alloys might be simpler than magmas, they were but little understood by chemists.

The remarkable series of conceptual developments of the 1880s that were associated with the genesis of physical chemistry supplied a basis for understanding that Bunsen and Durocher lacked (23). Van't Hoff's law of osmotic pressure, Arrhenius's theory of electrolytic dissociation, and the subsequent work of Nernst, Ostwald, and their students illuminated important aspects of the behavior and properties of solutions. Gibbs's phase rule, first appreciated and used by Roozeboom, Ostwald, Le Chatelier, and van't Hoff in the late 1880s, furnished scientists with a valuable guide to the study of heterogeneous equilibria. Although the laws of ideal solution were valid only for extremely dilute aqueous solutions -- slightly polluted water, as one chemist phrased it -- and although the phase rule was applied at first only to very simple systems, proponents of the new physical chemistry confidently asserted that their discoveries would serve as a basis for the reconstruction of chemistry and for the creation of new links between chemistry and other sciences (24). Physical chemistry, they suggested, would be a donor science, capable of stimulating development in, and answering the needs of, neighboring disciplines. It was such, they argued, because solutions were the sites of most chemical changes. The tests of the analytical chemist were conducted in the wet way, the phenomena of life took place in solution, solutions were the medium of many industrial processes, and last but not least, solutions were powerful agents of geological change. Insofar as the founders of physical chemistry had revolutionized the chemists' understanding of solutions, they had also transformed the basis for the geologists' understanding of the chemistry of rock formation and alteration (25, 26, 27).

Physical chemists were themselves among the first to appreciate the geological significance of their ideas and to apply their methods to geological problems. Both van't Hoff and Arrhenius, for instance, turned toward geological issues in the 1890s (25, 28). That physical chemists should have expressed a lively interest in geological phenomena is not surprising. They were anxious to develop such applications both as a means of legitimating their ideas and of strengthening the rather fragile institutional supports of their specialty. But a transmission of ideas depended as much upon the receptivity of geologists as on the assertiveness of physical chemists, and geologists generally were ill-prepared to follow, never mind make use of, recent developments in chemistry. Moreover, there was a great disparity between the accomplishments of the new physical chemistry and the needs of geologists. Magmas, for example, were far more complex than dilute aqueous solutions or alloys, and they were inaccessible to direct study. Whereas the chemist or metallurgist could check his conclusions against the behavior of solutions or alloys in the laboratory, the geologist did not possess facilities in which the natural conditions of rock formation could be duplicated. His givens were the rocks themselves, the final products of magmatic cooling. Their genesis was open to many interpretations.

The transmission of ideas and methods from physical chemistry to geology was therefore a slow and halting process, and the path and pace of the transmission varied a great deal from country to country. The geologists of France and England were rather slow to adopt a physico-chemical approach. Instead, the first geologists to make use of physical chemistry were located in countries within the orbit of German science: Norway (J. H. L. Vogt), Russia (A. Lagorio, F. Y. Loewinson-Lessing), Austria (C. A. S. Doelter), and the United States (J. P. Iddings, C. R. Van Hise, A. L. Day, N. L. Bowen) (29). Three factors appear to have been crucial in leading to this pattern: the presence of a vigorous group of petrographers interested in the problems of the classification and genesis of igneous rocks; the existence of a community of chemists acquainted with the recent developments in solution theory and chemical thermodynamics; and the existence or creation of institutions within which geologists and chemists might mix and collaborate. Where these conditions were present, geochemical research flourished at the turn of the century; where one or another of these conditions was absent, a geochemical tradition was slow to develop.

The importance of these conditions is well illustrated by the American case. By the mid-1890s, a large and still growing cadre of petrographers was active in America. Some had learned their science at home in the field, but at the core of this group were scientists who had acquired knowledge of the structure, texture, and composition of rocks in Germany. Beginning in the 1870s, a stream of American students had gone to Heidelberg, Leipzig, and other German universities to study under Ferdinand Zirkel, Harry Rosenbusch, and other German masters of petrographic techniques. A roster of the more notable of these Americans would include Joseph Paxton Iddings, Charles Whitman Cross, Henry S. Washington, Louis V. Pirsson, G. H. Williams, J. E. Wolff, F. E. Wright, and R. A. Daly. Although most went in order to learn the intricacies of classification and thin section analysis, many took advantage of their stay in Germany to attend lectures on chemistry as well (30-33).

After returning to the United States many began their careers at the U.S. Geological Survey. There, their primary task was to identify and map districts with economically interesting minerals, but they also had the freedom to study broader questions. The rocks of the eruptive districts of the American west, for instance, were an excellent site for study of the origins and differentiation of igneous rocks, and the ancient rocks of the Lake Superior region offered tempting materials for work on processes of metamorphism and ore formation. By the 1890s several American petrographers had begun to take advantage of these opportunities and to make significant contributions to the literature on the theoretical problems involved in the history and classification of rocks.

Foreign study among American students of petrography reached a peak in the early 1890s, and it was at this time that the first Americans began to return from Europe with advanced training in another specialty -- physical chemistry. Between 1889 and 1905, over forty Americans worked in Wilhelm Ostwald's laboratory at Leipzig; they formed the nucleus of a vigorous community of physical chemists in the United States. Like their European teachers, these physical chemists were anxious to win an audience for their ideas and to develop institutions for their specialty. During the 1890s, they translated European texts, established graduate programs, and launched a journal devoted to their subject. They also sought to adapt their knowledge to the needs of other specialists -- at first by writing texts on qualitative and quantitative analysis and general chemistry from the standpoint of the ionic theory, and later by applying solution theory and chemical thermodynamics to problems in agriculture, engineering, and medicine. During the nineties, they succeeded in bringing their subject to the attention of the broader scientific community and in establishing settings in which their successors might prosper (27).

Shortly after 1890, liasons began to be established between these two specialties. The first American contacts were not made in emulation of developments abroad; rather they were independent of, and roughly simultaneous with, similar contacts then being established between petrographers and physical chemists in Norway, Austria, and Russia. It was not the university that served as the principal context of intellectual commerce, but rather the offices and laboratories of the U.S. Geological Survey. And it was in the Survey and in an institution built by scientists from that agency -- the Geophysical Laboratory of the Carnegie Institution of Washington -- that the new physico-chemical geology flourished.

Two petrographers, Joseph Paxton Iddings (1857-1920) and Charles R. Van Hise (1857-1918) were the most influential of the intellectual brokers who established these contacts in America. They were not extremely original scientists nor were they laboratory men with training in physical chemistry. Rather they were synthesizers who cast their nets wide and who came to believe, through work in field and library, that physical chemists working in cooperation with geologists might develop answers to the perplexing problems of rock formation and metamorphism.

Iddings appears to have been the first American geologist to draw attention to the geological implications of the solution theory. He was one of the many American petrographers who had studied under Rosenbusch and who had then joined the young U.S. Geological Survey. During the 1880s, Iddings spent his summers in the field, amidst the volcanic rocks of Yellowstone Park, and his winters in Washington, where he struck up a friendship with F. W. Clarke, the chief chemist of the Survey and a man with wide knowledge of the new physical chemistry. Both of these experiences proved important for Iddings (30, 34).

His field work gave Iddings experience with eruptive rocks that exhibited a fairly smooth and continuous gradation of mineral and chemical composition; the deposits appeared to reflect their historical sequence of formation. Mineralogical and chemical analyses of samples from these beds, together with reports of similar deposits elsewhere, led Iddings to suggest that the igneous rocks of any region are so intimately connected by mineralogical and chemical relations that they must have originated in a common source -- some single magma that could give rise to various kinds of igneous rocks through a process of differentiation. Several European petrographers had already noted textural and structural evidence for the existence of discrete petrographic provinces, and others had suggested that igneous rocks might be derived from a single magma through differentiation. Iddings's contribution was to stress the chemical relatedness of rocks in particular provinces -- their chemical consanguinity (16).

Iddings's work led directly to two fundamental questions: What was the physical and chemical state of molten magmas? and by what process could a wide variety of distinct but chemically related rocks be derived from a single, homogeneous source? It was here that physical chemistry became important. In his first paper on the crystallization of igneous rocks, written in 1889, Iddings had been unable to go beyond his predecessors when it came to these problems. He treated magmas as saturated solutions of silicate molecules from which mineral species crystallized as temperature and pressure conditions changed. As to the exact nature of the compounds in solution and the details of differentiation in magmas, he confessed ignorance (35).

In his next paper on the subject, written in 1892, Iddings adopted a far more confident tone. Whereas he had earlier been without suggestions as to how to proceed, in this paper he advanced a new hypothesis regarding the condition of substances in magmas and posited a mechanism whereby chemical differentiation might occur. In large part this new confidence appears to have been derived from his discovery of the papers of van't Hoff and Arrhenius on solution theory. His attention was drawn to Arrhenius's work, he tells his reader, through the suggestion of of his colleague at the Survey, F. W. Clarke (36).

Iddings had already concluded that identical magmas could produce very different minerals depending upon the conditions of cooling. Analyses of rocks drawn from two locations in the Yellowstone region had shown that although these samples were chemically identical they differed markedly in mineral composition. Iddings had even tentatively suggested that minerals might not retain their molecular integrity in the molten state, but might decompose into simpler units which could shift about independently of one another and enter into several associations, depending upon the physical conditions prevailing during crystallization (37). Arrhenius's theory of electrolytic dissociation provided Iddings with a powerful support for this idea in that it indicated that magmas might be considered as belonging to a large class of solutions in which molecules of solute dissociated into ions. Iddings concluded that the simpler units in magmas were probably oxides of the constituent elements (36). Although this conclusion proved premature, Iddings's work was an important advance because it both put new content into Bunsen's old suggestion that magmas were solutions and went some way toward explaining how different rocks could arise from one magma. As to the mechanism of differentiation, Iddings once again drew upon the work of a physical chemist, in this case, van't Hoff. Van't Hoff's theory of osmotic pressure provided a theoretical basis for an empirical principle discovered in the early 1880s by the physicist, C. Soret: in a solution in which there is a temperature gradient, molecules of the solute will tend to concentrate in the cooler portion. This mechanism of molecular diffusion, Iddings suggested, might be sufficient to explain what he took to be evidence of differentiation in magmas prior to crystallization (38).

Iddings concluded his paper with an admonishment:

> The complexities of a compound solution that exists only at extremely high temperatures and experiences the pressures to which rock magmas have undoubtedly been subjected may long remain beyond the reach of direct investigation. Still the steady advance of experimental physics offers great possibilities in this direction. Until the establishment of definite knowledge concerning the nature of molten magmas we must proceed along the lines of analogy by applying to them such laws as may be found applicable to solutions that exist at lower temperatures and pressures (39).

One clear implication of Iddings's paper was that geologists should pay heed to developments in the borderland between chemistry and physics. They should make use of the theory of solution, and they should seek to foster research that would obviate the need to reason about magmas by analogy -- that is, research involving the direct study of molten silicates. Not long after this paper was written, Iddings's colleague, Charles R. Van Hise, came to the same conclusions by a somewhat different route.

Although Van Hise was an exact contemporary of Iddings, his education was, in most respects, inferior to that of his colleague. Van Hise did not study in Germany; his formal training was as a mechanical engineer at the University of Wisconsin. He learned his geology in the field, first as an assistant in the state survey of Wisconsin, where he became acquainted with the use of the microscope, and later as a geologist and division chief in the U.S. Geological Survey. Although Van Hise taught at his alma mater throughout the 1880s and 1890s, and later became its president, his professional life was oriented toward the Survey, where his primary responsibility was to study the iron and copper districts of the Lake Superior region (40, 41). A lesser intellect might have been satisfied to prepare useful but essentially descriptive reports on the topography and minerals of the area, but Van Hise had unusual discipline and tenacity. His descriptive work led him, during the early 1890s, to study the causes of ore deposition and the larger subject of metamorphism. Frustrated by his inadequate preparation in the physical sciences, Van Hise embarked on an ambitious campaign to educate himself in the principles of physics and chemistry. In 1902, he wrote:

> during the past five years, in order to handle the problems of geology before me, I have spent more time in trying to remedy my defective knowledge of physics and chemistry... than I have spent upon current papers in geology; and with, I believe, much more profit to my work(42).

Van Hise did not engage in this study alone. In the mid-nineties, using Survey funds, he hired a young physical chemist, A. T. Lincoln, to assist him with calculations and, it appears, to tutor him in the new literature (44, 45). The fruits of these labors appeared in a series of papers during the late 1890s and in his massive *Treatise on Metamorphism* of 1904, an encyclopedic work that some of his colleagues took to be the last word on the subject (45, 46).

Van Hise's treatment of metamorphism arose from the field observation that young rocks are often marked by numerous fissures and joints whereas older rocks show many signs of folding and flexure but few of fracture. He ascribed the difference to the physical conditions under which deformation occurred. Young rocks were deformed near the surface in what Van Hise called a zone of fracture. Here stress might result in rupture, but incumbent pressure was insufficient to cause fissures, once opened, to be closed. Ancient rocks were deformed beneath this zone, in a second region which Van Hise called the zone of flow. Here pressure was greater than the strength of any rock, and hence fissures were impossible. Deformation in this lower region resulted from the plastic flow of rocks.

Although Van Hise attributed the gross deformation of rocks to physical causes, he believed that their alteration was a chemical, or physico-chemical, problem. Thus, he maintained that a different set of chemical reactions characterized the alteration of rocks in each zone. In the zone of fracture, reactions typically occurred with the expansion of volume and the liberation of heat: oxidation, carbonation, and hydration. In the lower zone, these reactions were reversed as pressure rather than temperature became the factor controlling chemical change. In both zones, the alteration of rocks took place chiefly through the agency of water and the mineralizers that water carried in solution. In the zone of fracture, meteoric waters circulated through fissures, dissolving and depositing metals and other minerals as a result of variation in solubility arising from changes in termperature and pressure. In the zone of rock flow, minute quantities of trapped water acted to maintain rocks in a plastic state through the continuous solution and deposition of rock material (45, 47, 48, 49).

The names of van't Hoff, Arrhenius, Ostwald, and Nernst dot the pages of Van Hise's work and with good reason. His understanding of the effects of temperature and pressure on chemical reactions and of the roles of water and ionic equilibria in metamorphic processes was derived largely from his reading of the work of these physical chemists. "The handling of the problems of rock alteration with fairly satisfactory results," he later wrote, "was possible because of the rise of physical chemistry. Had this science not been developed within the past score of years, it would not have been possible to have gone far upon the problem...." (50). Van Hise thought that he had sketched out the major features of a general theory of metamorphism. But he recognized that the details of metamorphic processes were largely unresolved. A better understanding of the phenomena, he suggested on more than one occasion, would require experimental study by scientists with a surer knowledge of physical chemistry than he had been able to acquire (51, 52).

Despite differences in their training and experience, Van Hise and Iddings both came to the view that a closer collaboration between chemists and geologists would be necessary, and both perceived advantages in conducting such collaborative work in the laboratory. But formidable obstacles confronted such an undertaking. Facilities did not exist in the United States or abroad for approximating and controlling the conditions under which rocks were formed (53, 54). Advances in the technology of producing and measuring high temperatures and pressures suggested that such conditions might be realized, but a laboratory equipped to study molten silicate solutions would be an expensive affair, requiring air and water compressors, high pressure bombs, gas and resistance furnaces, costly electrical equipment, and prodigious quantities of platinum for reaction vessels. A fairly large staff with a variety of special skills would be necessary to design and supervise delicate experiments that might run for hours or days. And the entire enterprise would be directed toward assembling data with no demonstrable economic benefit. Although Van Hise was a professor at Wisconsin and Iddings had joined the staff at the University of Chicago in 1892, there is no evidence that either approached university officials with such a project. Rather they chose to work through the U.S. Geological Survey.

The opportunity came in 1900, when the director of the Survey, C. D. Walcott, appointed Van Hise chairman of a committee to study relations between the geological and chemical divisions of the agency (55). The Survey had sponsored a limited program of geophysical investigations during the 1880s and early 1890s, but during the budget crisis of 1892, this laboratory had been abolished and the Survey had confined its experimental studies to routine chemical analyses (56). Now, however, the chemists in the Survey, together with some of the laboratory minded geologists such as Van Hise, had grown restive with this role. The report of Van Hise's committee, not surprisingly, recommended that chemical research in the Survey be greatly expanded and that time and equipment be made available for research on problems such as had caught the attention of Van Hise and Iddings (57).

Later that year, a laboratory of chemistry and physics was organized within the Survey under the direction of George F. Becker, a scientist with both laboratory and field experience (58, 59). It became the kernel from which a geochemical tradition grew in America. Becker drew together an impressive staff whose members had backgrounds in several disciplines: the analytical chemists, F. W. Clarke and W. F. Hillebrand; Arthur Louis Day, a Yale trained physicist who had worked on high temperature thermometry at the Physikalisch-technische Reichsanstalt; E. T. Allen, a young chemist with considerable interest in the new physical chemistry; F. E. Wright, a petrographer fresh from Rosenbusch's institute; J. K. Clement, a physical chemist trained by Nernst at Göttingen; and another physical chemist, E. S. Shepherd, who had worked at Cornell with W. D. Bancroft, the leading American expert on phase equilibria (60).

This multi-disciplinary staff developed an ambitious interdisciplinary program for the experimental study of rock formation. "Our plan," Day and Allen wrote,

> was to study the thermal behavior of some of the simple rock making minerals by a trust-worthy method, then the conditions of equilibrium for simple combinations of these, and thus to reach

> a sound basis for the study of rock formation or differentiation from magma. Eventually, when we are able to vary the pressure with the temperature over considerable ranges, our knowledge of the rock-forming minerals should become sufficient to enable us to classify many of the earth-making processes in their proper place with the quantitative physico-chemical reactions of the laboratory (61).

Part of this program was realized at the Survey. Between 1901 and 1907, the chemist Allen, the physicist Day, and the petrographer, Iddings, completed a rigorous study of the thermal properties of the plagioclase feldspars, the most abundant rock-forming minerals. This study was important in several respects. It was the first fully quantitative study of the behavior of cooling silicate melts. Its authors introduced many of the techniques that later became standard features in such research, for instance, the use of thermal rather than optical methods to determine melting points and the use of artificial minerals to guarantee purity. Moreover, Day and Allen explicitly tied their findings to the work of the Dutch school of physical chemists by showing that sodium and calcium feldspars formed a continuous series of isomorphic crystals that fit Type I of Roozeboom's classification of heterogeneous equilibria. This result cast light both on the process of differentiation -- ruling out, for example, the idea that all silicate melts behaved as eutectics -- and it also allowed Day and Allen to account for one of the characteristic features of feldspar crystals, their zonal structure. Walcott called this work "one of the most important contributions to geologic physics ever printed"; it was a model for subsequent studies of other silicate systems (62).

The research program, begun at the U.S. Geological Survey, flourished at a new institution that opened its doors in 1907, the Geophysical Laboratory of the Carnegie Institution of Washington. Several features of the history of this Laboratory merit emphasis here. During its planning stage, which lasted from 1902 to 1905, Van Hise, Iddings, and other American petrographers played a crucial role in guiding the trustees of the Carnegie Institution of Washington toward funding the physico-chemical study of rock formation. When it opened, the Geophysical Laboratory was based largely upon the staff, techniques, and research program that had evolved at the Geological Survey. Arthur L. Day was its first director; Allen his chief chemist. The physical chemists, Clement and Shepherd moved with them, as did the petrographer, Wright. Using this group as a nucleus, Day expanded the staff by hiring other petrographers and physical chemists: students of Rosenbusch and Zirkel and former pupils of the American physical chemists, A. A. Noyes and W. D. Bancroft. Following the procedures established at the U.S.G.S., these scientists went about the task of methodically collecting the data necessary to construct phase diagrams of two and three component mineral systems. The results were of cardinal importance to geochemistry. Their data went a long way toward filling a vacuum in the chemist's knowledge of the behavior of silicates and served as the basis for N. L. Bowen's theory of the evolution of igneous rocks, which, although not unchallenged, has remained a cornerstone of teaching and research in petrology and geochemistry since it was advanced in 1922 (56, 63, 64).

The influence of the Geophysical Laboratory as a stimulus toward geochemical research and training was perhaps as important as the work conducted within its walls. When the Laboratory was founded, neither its promoters nor its sponsor envisioned its work as having any immediate economic importance. Rather it was established as a center for basic research into the processes of petrogenesis. But soon after the Laboratory opened, its staff members began to make contributions of industrial significance -- spinoffs from their basic research program. Day and Shepherd developed a practical method for the production of quartz glass; others made discoveries of value to the cement and ceramics industries (65-68). During World War I, the staff was responsible for developing the methods that allowed the United States to become a major producer of optical glass (69). After the war, many of the staff members found positions in industrial research as managers of firms such as U.S. Steel, Corning Glass, and Pittsburgh Plate Glass, and realized that the techniques used in the study of minerals could also be applied to glass, ceramics, cement, and metals. And as the market for geochemists' knowledge began to develop, so too did interest in the field in universities. Other staff members left the Geophysical Laboratory after World War I to take positions at Harvard, Yale, Chicago, and other centers where they established some of the first graduate courses and research facilities in the fields of physico-chemical petrology and geochemistry (56). Nor were these developments neglected abroad. During the teens and twenties, a stream of foreign scientists visited Washington to study the methods of Day, Bowen, and their associates. The geophysical institutes in Sendai, Japan and Zurick, Switzerland were founded by scientists who had first worked at the Geophysical Laboratory; visitors also came from Oslo, Groningen, and Amsterdam (70).

It would be no exaggeration to say that the Geophysical Laboratory was at the hub of developments in geochemistry in the early twentieth century. It was a site where physical chemists and earth scientists met, exchanged ideas, and carried the physico-chemical study of petrogenesis toward its limits in a laboratory setting. It was also a center that made vital contributions both to the formation of an international community of geochemical investigators and to the development of markets for their services.

By 1915, geochemistry had become one of those fields in which American scientists had begun to assume international leadership. Americans in this field went to Europe not so much to learn as to teach; Europeans visited America not so much to instruct as to study. The sciences upon which geochemistry was based, petrography and physical chemistry, both had their origins in Europe, but Americans such as Iddings, Van Hise, Day, and Bowen had been in the vanguard of those who sought to bring them together.

Geochemistry was not, of course, the only scientific specialty in which Americans were acquiring reputations for excellence in the early twentieth century. Similar advances were being made in other fields, and especially in new specialties: genetics, physical chemistry, the biomedical and engineering sciences. In seeking to explain the success of American workers in these sciences it has become customary to look toward special features of the American market or American university. Market explanations generally account for differences between the rate of development of specialties in different national contexts through reference to local demands for skills and services (4). Explanations based on the structure of American institutions of higher learning stress the flexibility of the department and elective system and the adaptability of a decentralized and competitive network of universities (6, 7, 71).

Neither of these sorts of explanations, however, seems relevant in accounting for why the United States should have become a leading site for research in geochemistry at the turn of the century. When Iddings and Van Hise began in their rather speculative way to apply physical chemistry to geological problems they held no hope of immediate and direct practical consequences. When Day embarked on the planning of the Geophysical Laboratory, he appears to have envisioned little if any economic benefit. These workers did not respond to market demands, but rather were motivated by the desire to account for the origins, differentiation, and alteration of rocks -- questions that were of considerable importance to the history of geological structures and the classification of rocks, but which could hardly be claimed as problems of economic geology. The applications that later arose and contributed toward creating a market for geochemists' skills were largely, if not wholly, unanticipated (56).

Nor is it possible to invoke the dynamism of the American university and the flexibility of its various structural features as an explanation of the genesis of an American tradition in geochemistry. The university was very much on the periphery of this story. It supplied training to petrographers and physical chemists who worked at the U.S. Geological Survey and the Geophysical Laboratory, but courses and opportunities for geochemical research in American universities developed very slowly. When Day selected the staff of the Geophysical Laboratory, he picked scientists with credentials in petrography and physical chemistry; they were molded into geochemists after entering the Laboratory. Universities such as Chicago and Columbia became important sites for geochemical research and training only after 1930, and largely under the stimulation of developments taking place outside of the university setting. In the case of geochemistry, the university was not the matrix of specialization. Rather it was the U.S. Geological Survey and later the Geophysical Laboratory.

This is not to deny the importance of markets and universities in the genesis and development of new research fields in turn of the century America. It is to suggest however that we should beware of laying undue stress on market forces and institutional structures at the expense of intellectual factors. There is an intellectual basis for specialization. Scientists are not powerless captives of market conditions nor is their science a passive stuff that adapts itself to the configuration of existing institutions. Scientists are, or can be, active agents who shape the vessels in which they work in accordance with their perceptions of conceptual needs and opportunities. In the case of geochemistry, the concepts were drawn largely from petrography and physical chemistry. They came together in America because there were representatives of both of these traditions present who perceived a fit between the problems of the one field and the techniques of the other and who were in positions to mold institutions within which a new specialty might grow. A certain degree of institutional flexibility was prerequisite, but in this case, that flexibility was not to be found in the university but rather in the much maligned Geological Survey of the post-Powell era and in the Carnegie Institution of Washington.

Acknowledgments

I wish to thank Dr. Larry Owens of Princeton University and Dr. R. E. Gibson of the Applied Physics Laboratory of the Johns Hopkins University for their comments on an earlier draft of this essay, and I gratefully acknowledge the support of the American Council of Learned Societies and the National Science Foundation.

Literature Cited

1. Kevles, Daniel J. "The Physicists: The History of a Scientific Community in Modern America"; Knopf: New York, 1978.
2. Badash, Lawrence. "Radioactivity in America: Growth and Decay of a Science"; Johns Hopkins University Press: Baltimore, 1979.
3. Rosenberg, Charles E. "No Other Gods: On Science and American Social Thought"; Johns Hopkins University Press; Baltimore, 1976; Part 2.
4. Kohler, Robert E. "From Medical Chemistry to Biochemistry: The Making of a Biomedical Discipline"; Cambridge University Press: Cambridge, 1982.
5. Veysey, Laurence R. "The Emergence of the American University"; University of Chicago Press: Chicago, 1975.

6. Ben David, Joseph. "The Scientist's Role in Society: A Comparative Study"; Prentice-Hall: Englewood Cliffs, N.J., 1971; pp. 139-168.
7. Dolby, R. G. A. Annals Sci. 1977, 34, 287-310.
8. Higham, John. In "The Organization of Science in Modern America": Oleson, Alexandra and Voss, John, Ed.; Johns Hopkins University Press: Baltimore, 1979; pp. 3-18.
9. Shils, Edward. In "The Organization of Science in Modern America" (Ref. 8); pp. 19-47.
10. Manten, A. A. Chem. Geol. 1966, 1, 5-31.
11. Loewinson-Lessing, F. Y. "A Historical Survey of Petrology"; Tomkeieff, S. I., Trans.; Oliver & Boyd: Edinburgh, 1954.
12. Brock, W. H. In "Images of the Earth: Essays in the History of the Environmental Sciences": Jordanova, L. J. and Porter, Roy S., Ed.; British Society for the History of Science: Chalfont St. Gilles, 1979; pp. 147-170.
13. Allen, David Elliston, In "Images of the Earth" (Ref. 12); pp. 200-212.
14. Ref. 11, pp. 1-7 and 33-35.
15. Cross, Charles Whitman. Journ. Geol. 1902, 10, 322-376 and 451-499.
16. Iddings, Joseph Paxton. Bull. Phil. Soc. Wash. 1892-1894, 12, 89-214, esp. 91-127.
17. Bascom, F. Johns Hopkins Studies in Geology 1927, no. 8, 33-82.
18. Knopf, Adolph. In "Geology, 1888-1938. Fiftieth Anniversary Volume of the Geological Society of America"; Geological Society of America, 1941; pp. 335-363.
19. Larsen, E. S. In "Geology, 1888-1938" (Ref. 18); pp. 393-413.
20. Bunsen, Robert. Pogg. Ann. Phys. u. Chem. 1851, 83, 197-272.
21. Bunsen, Robert. Zeitsch. Deutsch, Geol. Ges. 1861, 13, 61-63.
22. Durocher, J. Ann. Mines 1857, 11, 220.
23. On the origins of physical chemistry see Root-Bernstein, Robert Scott. Ph.D., Thesis, Princeton University, Princeton, N.J., 1980.
24. Bancroft, W. D. In "A Half-century of Chemistry in America, 1876-1926"; Browne, Charles A., Ed.; Chemical Publishing Co.: Easton, Pa., 1926; p. 94.
25. Hoff, J. H. van't. "Physical Chemistry in the Service of the Sciences"; University of Chicago Press: Chicago, 1903.
26. Jones, Harry Clary. "A New Era in Chemistry"; Constable & Co.: London, 1913.
27. Servos, John W. Ph.D., Thesis, The Johns Hopkins University, Baltimore, Md., 1979; Chapter 2.
28. Riesenfeld, Ernest H. "Svante Arrhenius"; Akademische Verlagsgesellschaft: Leipzig, 1931; pp. 40-47.
29. Ref. 17, pp. 44ff.
30. Mathews, E. B. Bull. Geol. Soc. Am. 1933, 44, 352-374.
31. Larsen, E. S. Biogr. Mem. N. A. S. 1958, 32, 100-112.

32. Knopf, Adolph. Biogr. Mem. N. A. S. 1960, 34, 228-248.
33. Birch, Francis. Biogr. Mem. N. A. S. 1960, 34, 30-64.
34. Dennis, L. M. Biogr. Mem. N. A. S. 1934, 15, 139-165.
35. Iddings. Joseph Paxton. Bull. Phil. Soc. Wash. 1888-1891, 11, 65-114.
36. Ref. 16, pp. 154-156.
37. Iddings, Joseph Paxton. Bull. Phil. Soc. Wash. 1888-1891, 11, 191-220, esp. 212-213 and 217.
38. Ref. 16, pp. 158-160.
39. Ref. 16, p. 194.
40. Chamberlin, T. C. Journ. Geol. 1918, 26, 690-697.
41. Vance, Maurice. "Charles Richard Van Hise: Scientist Progressive"; State Historical Society of Wisconsin: Madison, 1960.
42. Van Hise, Charles R. Science 1902, 16, 326.
43. A. T. Lincoln to C. R. Van Hise, 19 October 1899, U. S. G. S. Lake Superior Division Official Correspondence, Box 6, State Historical Society of Wisconsin.
44. C. R. Van Hise to A. T. Lincoln, 24 October 1899, U. S. G. S. Lake Superior Division Official Correspondence, Van Hise Letterbooks, State Historical Society of Wisconsin.
45. Van Hise, Charles R. "A Treatise on Metamorphism"; U. S. Government Printing Office: Washington, 1904.
46. Miller, Benjamin LeRoy. Johns Hopkins Studies in Geology 1927, no. 8, 121-135.
47. Van Hise, Charles R. Journ. Geol. 1900, 8, 730-770.
48. Van Hise, Charles R. Bull. Geol. Soc. Am. 1898, 9, 269-328.
49. Van Hise, Charles R. Journ. Geol. 1904, 12, 589-616.
50. Ref. 49, p. 609.
51. Ref. 49, pp. 610-611
52. Ref. 42, pp. 322-323.
53. Ref. 19, pp. 395-398.
54. Ref. 11, pp. 42-48.
55. C. D. Walcott to C. R. Van Hise, 24 February 1900, U. S. G. S. Lake Superior Division Correspondence, Box 6, State Historical Society of Wisconsin.
56. Servos, John W. Hist. Stud. Phys. Sci., in press.
57. C. R. Van Hise, J. S. Diller, and S. F. Emmons to C. D. Walcott, 13 March 1900, Box 17, George F. Becker Papers, Library of Congress.
58. Merrill, George. Biogr. Mem. N. A. S. 1920, 21, 1-13.
59. Day, Arthur L. Bull. Geol. Soc. Am. 1919, 31, 14-19.
60. The work of this laboratory may be followed in the U. S. Geological Survey, Annual Report.
61. Day, A. L. and Allen, E. T. "Isomorphism and the Thermal Properties of the Feldspars"; Carnegie Institution of Washington: Washington, 1905; p. 17.
62. U. S. Geological Survey, Annual Report 1904-1905, 26, 99.
63. Eugster, Hans P. Biogr. Mem. N. A. S. 1980, 52, 34-79.
64. Tilley, C. E. Biogr. Mem. F. R. S. 1957, 3, 6-22.

65. Day, A. L. and Shepherd, E. S. Science 1906, 23, 670-672.
66. Shepherd, E. S. and Rankin, G. A. Am. Journ. Sci. 1909, 28, 292-333.
67. Bates, P. H. Journ. Franklin Inst. 1922, 193, 293-294.
68. Sosman, Robert B. "The Properties of Silica"; Chemical Catalog Co.: New York, 1927; pp. 809-812.
69. Wright, F. E. "The Manufacture of Optical Glass and of Optical Systems: A Wartime Problem"; U. S. Government Printing Office: Washington, 1921; pp. 10-13.
70. Arthur L. Day to John C. Merriam, 26 July 1930, Geophysical Laboratory Projects--High Pressure Research Program, files of the Carnegie Institution of Washington, Washington, D. C.
71. The two forms of explanation are not mutually exclusive. See Guedon, Jean-Claude. In "History of Chemical Engineering": Furter, William F., Ed.; American Chemical Society: Washington, 1980; pp. 45-75.

RECEIVED November 10, 1982

2

Synthetic Petroleum from High-Pressure Coal Hydrogenation

ANTHONY N. STRANGES

Texas A&M University, History Department, College Station, TX 77843

This paper examines the history of high-pressure coal hydrogenation from its beginning in pre-World War I Germany to its present state of development in the United States. It shows how Germany, a country almost entirely dependent on imported natural petroleum for its liquid fuel requirements, called on science and technology to make it independent of natural petroleum. Recognizing petroleum as the fuel of the future, her scientists took coal, a substance of which they had an abundance, and from it synthesized petroleum, a product of which Germany had none.

Of the several processes used to convert coal into petroleum, high-pressure coal hydrogenation was the most highly developed. Its history falls into three broad periods: (1) the work of Friedrich Bergius, the inventor of high-pressure coal hydrogenation, in the period of 1910-1925; (2) the commercial development of Bergius's process by German industrialists in the 1920s, 1930s, and 1940s and (3) the present state of coal hydrogenation in the United States.

Before the rise of Germany's synthetic petroleum industry in the mid-thirties, a shortage of liquid fuel seriously threatened her economic and social well-being. If such a situation should repeat itself in the United States, a synthetic fuels program could contribute towards American self-sufficiency in energy.

The availability of natural resources has for the most part established the history of energy consumption in the United States. This history includes four broad time spans: (1) the wood period,

0097–6156/83/0228–0021 $06.25/0

(2) the coal age, (3) the petroleum era, and (4) the very recent epoch of uncertainty. The first or wood period began with the American Revolution and reached its peak in 1850 when wood supplied 90 percent of the energy consumed in the United States. Wood consumption decreased rather regularly after that, falling to 50 percent in 1885 and to only 5 percent today.

The coal age began in the second half of the nineteenth century, when coal consumption increased steadily. In 1850 coal supplied 10 percent of the total energy consumed but by 1885 this rose to 50 percent. Coal and wood were at that time the United States's major energy sources. While the use of wood as an energy source continued to decrease in the nineteenth century, coal consumption attained a maximum in 1910, accounting for 80 percent of the energy expended in the United States. After 1910, coal consumption started to decline, dropping to approximately 45 percent of the total energy consumed in 1945 and to 20 percent today.

Edwin Drake made the first significant petroleum strike in 1859 but the petroleum era really did not begin until the twentieth century. Petroleum, which provided 5 percent of the United States's energy demands in 1890, was with natural gas the source of 45 percent of energy consumed in 1945. Consumption has increased steadily and has fallen only very recently. (1) This leads to the fourth time span, the epoch of uncertainty, which began with the Arab oil embargo in the winter of 1973-1974. It includes wood, coal, and petroleum among its energy sources plus such alternate energies as wind, nuclear, solar, biomass, geothermal and the synthesis of petroleum from coal.

The synthesis of petroleum has received much attention lately. Indeed, many scientists believe the synthetic petroleum obtained from coal hydrogenation will contribute significantly to the United States's natural petroleum supply. Coal hydrogenation is not a recent discovery; it has a long and successful history dating from 1913.

Early Research on Coal Hydrogenation

Coal hydrogenation is a process which converts different varieties of coal into synthetic petroleum by reacting coal with hydrogen gas at high pressure and high temperature. A German chemist, Friedrich Bergius (1884-1949) was the first to hydrogenate coal successfully and to foresee its commercial development. Due largely to the mass production of automobiles, worldwide petroleum consumption had increased steadily in the first decade of the twentieth century. The United States, the world's leading petroleum producer, needed forty-one years from its first substantial strike in August 1859 to December 1900 to extract its first billion barrels of petroleum, but only eight years for its next billion. (2) For Germany, a country with tremendous coal deposits but little natural petroleum, Bergius saw in coal hydrogenation a way to eliminate Germany's almost total dependence on imported petroleum to meet her ever-increasing petroleum requirements. (3)

High-pressure research was a relatively new field of investigation and it appeared very promising to many chemists in the first decade of the twentieth century. (4) To acquire high-pressure research experience, Bergius spent a year working with two of its leaders, Walter Nernst (1864-1941) in Berlin and Fritz Haber (1868-1934) in Karlsruhe, after obtaining his doctorate in 1907. Bergius then moved to a small factory in Hanover and in 1910 began independent high-pressure research. His first experiments were on the hydrogenation of an artificial coal he had prepared from cellulose, (5) followed by his most significant work, the hydrogenation of natural coals.

By summer of 1913, Bergius had successfully hydrogenated natural coals and showed for the first time that the reaction produced synthetic petroleum. He carried out these early experiments in a relatively small steel autoclave (a high-pressure container) of 400 ℓ capacity, treating 150 kg (330 lb) batches of powdered coal with five kg of hydrogen gas for twelve hours at 400°C and 200 atm pressure. Of all the coals tested, Bergius found that the younger coals, brown coals, lignites, and bituminous, gave the best yield, about 85 percent conversion to synthetic petroleum. He showed that they had a carbon content of about 85 percent, and that the high-pressure reaction raised their hydrogen content from 6 to nearly 15 percent. This was the key to Bergius's coal liquefaction process. He could hydrogenate coals with less than 85 percent carbon but not those whose carbon content exceeded 85 percent. (6)

Bergius also established that coal liquefaction consisted of two competing reactions: (1) hydrogen addition at temperatures and pressure of 300-400°C and 200 atm respectively; this changed the complex solid coal structure into heavy pitch-like hydrocarbons and (2) splitting the heavy hydrocarbons into lighter molecules which converted the pitch into liquid hydrocarbons at about 450°C. In other words, the hydrogenation of coal was a cracking process that absorbed hydrogen. In the autumn of 1913, Bergius took out the first patent on coal hydrogenation. (7)

The Moves to Essen and Rheinau: World War I

By 1914, Bergius was ready to enlarge the coal hydrogenation process and he accepted an offer to move from his small factory in Hanover to the Essen Works of Theodore Goldschmidt A.G., a petroleum refining company. At that time the only industrial-size, high-pressure apparatus in use was at the Badische Anilin- und Soda-Fabrik's synthetic ammonia plant in Ludwigshafen-Oppau. Bergius planned to construct in Essen industrial-size apparatus for the more difficult hydrogenation of a solid coal with hydrogen gas. He intended to make the entire operation--the feeding, mixing, reaction, and removal of products--a continuous process. (8)

World War I, which began on August 1, 1914, revealed Germany's critical need for petroleum. The German General Staff had counted on a quick military victory but within a short time they suddenly realized that the war was going to last longer than expected. Their commitment to a swift victory was the basic war philosophy of the late Count Alfried von Schlieffen (1833-1913), chief of the General Staff from 1895 to 1906. The Schlieffen Plan was the logical outcome of the quick victory over France in 1870-1871, and from that time the General Staff had reworked and refined it for the war Germany was presently fighting against Britain, France, and Russia. To carry out the plan, the vast bulk of the German army would launch a tremendous assault on France, while a minimal force held off the Russians. With the defeat of France, Russia would then easily fall before the German army's full force. England, now alone, would have no choice but to surrender if she extended the war. (9)

The Schlieffen Plan almost worked. But at the Battle of the Marne during the second week of September 1914, an unexpected French counterattack stopped cold the German army's rapid advance to Paris. German industry up to this time played no part in the Schlieffen Plan. Now, with all hope ended for a quick victory, the German High Command recognized the grim prospects of fighting a war with limited supplies of such strategic materials as sodium nitrate and petroleum.

Karl Bosch (1874-1940) and Alwin Mittasch (1869-1953) of Badische Anilin- und Soda-Fabrik eliminated the nitrate shortage that occurred after the British sea blockade effectively cut off the nitrate supply from Chile. By May of 1915, they had successfully developed at their Oppau Plant an industrial-scale process for oxidizing ammonia. Their process converted the large quantities of synthetic ammonia produced by the Haber process to nitric acid and other nitrates that were essential for fertilizers and explosives. (10)

Bergius's efforts to establish an industrial-scale coal hydrogenation process for the production of synthetic petroleum were not as successful. The Konsortium für Kohlenchemie in 1916 gave him 30 million marks to construct a large factory at Rheinau near Mannheim but Bergius never solved the major operating problems until long after the war had ended. Development also languished, particularly during the later stages of the war, because Germany had obtained access to Rumanian oilfields. The conversion of coal to synthetic petroleum thus became of minor importance. (11) Indeed, Bergius did not resume active research at Rheinau until 1921, construction of the factory taking place only in 1924.

The major operating difficulties that confronted Bergius's promising research on coal hydrogenation were not entirely technical, however. Obtaining sufficient research funds proved a serious problem in postwar, inflation-plagued Germany. To secure the necessary funds, Bergius in 1921 formed the International Bergius Company, turning over to it effective control of the Rheinau plant and all foreign rights to his hydrogenation process. German and Dutch investors divided equally the capital of the new company. (12) That same year Bergius organized a British Bergius Syndicate to experiment with British coals, and if the hydrogenation was successful, to develop the process within the British Empire. (13)

Over a three-year period at Rheinau, 1922-1925, Bergius and his assistants tested more than 200 different kinds of coal. Starting from a relatively small scale, they eventually hydrogenated coal in quantities as large as 1,000 kg (1 ton). A typical reaction run contained 100 kg of coal mixed with 40 kg of heavy oil, 5 kg of hydrogen gas, and 5 kg of ferric oxide to remove any sulfur present in the coal. The reaction yielded 20 kg of gas and about 128 kg of oil and solids. Distillation of the oil produced 20 kg of gasoline. (14)

Expanding the coal hydrogenation process to a small factory-size operation at this time forced Bergius to seek a larger and more economical source of hydrogen gas than he had obtained from the iron-superheated water reaction. (15)

$$\underset{\text{Iron}}{3Fe} + \underset{\text{Steam}}{4H_2O} \rightarrow \underset{\text{Iron Oxide}}{Fe_3O_4} + \underset{\text{Hydrogen}}{4H_2}$$

As Bergius pointed out, the economic success of coal hydrogenation largely depended on a cheap and convenient hydrogen supply. This led him to try the well-known water gas reaction, but using as starting materials methane and ethane gas which were present to a considerable extent in the gaseous products of coal hydrogenation. (16)

$$\underset{\text{Methane}}{CH_4} + \underset{\text{Steam}}{H_2O} \rightarrow \underset{\text{Carbon Monoxide}}{CO} + \underset{\text{Hydrogen}}{3H_2}$$

$$\underset{\text{Ethane}}{C_2H_5} + \underset{\text{Steam}}{2H_2O} \rightarrow \underset{\text{Carbon Monoxide}}{2CO} + \underset{\text{Hydrogen}}{5H_2}$$

The I.G. Farben Process and Its Commercialization in Germany

In 1927 Bergius brought his coal hydrogenation program to a successful conclusion. In the Rheinau-Mannheim factory, where he spent millions of marks in research and employed 150 men, Bergius had demonstrated the commercial potential of coal hydrogenation. But two problems remained. First, Bergius did not study the influence of different catalysts on the reaction. Secondly, his process was a one-stage operation: the hydrogenation and decomposition of the coal into synthetic petroleum took place in one step and resulted in a smaller yield of gasoline. The gasoline was of low quality and unless refined it could not compete with gasoline obtained from natural petroleum. (17)

The Badische Anilin- und Soda-Fabrik in Leuna, later a subsidiary of the German industrial giant, I.G. Farben, solved the problems Bergius encountered at Rheinau after reaching an agreement with him in 1924 to continue the research on coal hydrogenation. Within two years, in 1925, Matthias Pier (1882-1965), Badische's director of research, succeeded in separating Bergius's one-stage operation into two: hydrogenation of coal to a heavy oil in the first or liquid phase, decomposition of the heavy oil to gasoline-size molecules in the second or vapor phase. (18) Pier also directed the successful search for the catalysts that accelerated each phase of the hydrogenation process, discovering that metallic sulfides were insensitive to sulfur, usually one of the worst catalyst poisons and almost always present in coal. (19)

In I.G. Farben's two-stage process, the liquid phase required grinding the coal, suspending it in a heavy oil and adding a catalyst, heating the coal paste to 300°C, and then pumping it into convertors where the paste combined with hydrogen gas at 230 atm pressure. The liquid phase products consisted of 10 to 20 percent gases, 5 to 10 percent solids, 50 to 55 percent heavy oil and 20 to 30 percent middle oils such as kerosene, diesel oil, and heating oil. Molybdenum catalysts performed very well in the liquid phase hydrogenation but their limited supply compelled I.G. Farben to use low cost iron catalysts. In the vapor phase, the middle oils produced in the liquid phase reacted with hydrogen gas over fixed-bed catalysts at a temperature and pressure of 400°C and 200-300 atm. The catalysts were usually tungsten sulfide carried on terrana earth or activated alumina. The second stage hydrogenation yielded 50-70 percent gasoline with a boiling range to 165°C. (20)

From the first coal hydrogenation plant constructed in April 1927 at Leuna, which by 1931 had the capacity to produce 300,000 metric tons (2.5 million barrels) of synthetic petroleum per year, the German coal hydrogenation industry grew to twelve large plants (Böhlen, Scholven, Madgeburg, Welheim and others) in 1944. (21) Karl Bosch, I.G. Farben's chief executive, was primarily responsible for Germany's extensive development of the Bergius high-pressure liquefaction of coal rather than a second coal conversion process, Franz Fischer and Hans Tropsch's indirect synthesis of petroleum from carbon monoxide and hydrogen. (22) Bosch had successfully transformed Fritz Haber's ammonia synthesis from a laboratory process to a commerical industry and vowed to do the same for coal hydrogenation. At their peak in 1944, the German coal hydrogenation plants produced over three million metric tons (25.5 million barrels) of synthetic petroleum of which two million metric tons (17 million barrels), after adding lead, were high quality aviation and motor gasoline of 100 octane number. In World War II these plants provided most of the German military's fuel. (23)

The cost of hydrogenating coal was high, 190 marks per ton, or the equivalent of 24 cents per gallon. (24) This was more than double the price of imported gasoline, but for Germany, with only a limited supply of natural petroleum, no alternative remained during the War other than the construction of synthetic petroleum plants. In this way Germany utilized her naturally abundant supplies of bituminous and brown coal.

The German Energy Plan and Its Results

Germany's synthetic fuel industry never could have developed without the government's support. As an incentive toward the production of synthetic petroleum, the German government enacted a high tariff on imported petroleum in the early 1930s. It also entered into partnership agreements with German industry that were virtually risk-free to the industry.

In the Fuel Agreement (Benzinvertrag) of December 14, 1933, between I.G. Farben and the Reichswirtschaftsministerium (Office of the Secretary for Economy), I.G. Farben promised to produce at least 2,490,000 barrels of synthetic gasoline per year by the end of 1935 and to maintain this production rate until 1944. It set the production cost, which included depreciation, five percent interest, and a small profit, at 18.5 Pfennig per liter (1.1 quarts). (25) The government not only agreed to support this price but to pay I.G. Farben the difference between the production cost and any possibly lower market price, and to buy the gasoline if no other market emerged. On the other hand, I.G. Farben had to pay the government the difference between the 18.5 Pfennig per liter, which was at that time more than three times the world market price, and any higher price obtained on the market. By 1944, I.G. Farben had paid 85 million Reichsmark to the German government. (26)

In September 1936, Adolf Hitler announced his Four Year Plan to make the German military ready for war in four years and the German economy independent and strong enough to maintain a major war effort. Since Hitler's war strategy required petroleum, the development of a petroleum-independent Germany became the Four Year Plan's major thrust. Indeed, in 1936, Hitler urged the petroleum and the rubber industries to become independent of foreign production in eighteen months. (27)

Germany's synthetic petroleum industry never reached these goals, but production increased dramatically under the Four Year Plan. In 1933, only three small synthetic petroleum plants (Ludwigshafen-Oppau, Leuna, Ruhrchemie-Sterkrade-Holten) were operating, the last a Fischer-Tropsch plant. At that time, Germany's petroleum consumption was about one-half of Great Britain's, one-fourth of Russia's, and one-twentieth that of the

United States. Yet, even at such low consumption, domestic resources were inadequate; Germany imported 85 percent of her petroleum. By 1939, fifteen synthetic petroleum plants were in operation. In 1944, twenty-two coal hydrogenation and Fischer-Tropsch plants converted coal into gasoline and other petroleum products.

Clearly, Germany had the first successful synthetic petroleum industry producing 128 million barrels in the period 1938-1945. She did not continue work on coal liquefaction after World War II because the Potsdam (Babelsberg) Conference of July 16, 1945 prohibited it. (28) The Frankfort (Frankfurt) Agreement, four years later on April 14, 1949, ordered dismantling of all coal conversion plants, (29) but a new agreement, the Petersberg (Bonn) Agreement of November 22, 1949, quickly halted the dismantling process. (30) The West German government completely removed the ban on coal hydrogenation in 1951, though by this time the plants in West Germany, after design modification, were refining natural petroleum rather than hydrogenating coal. (31)

The Russians dismantled four of the hydrogenation plants located in their zone at Pölitz, Madgeburg, Blechhammer, and Auschwitz, and reassembled them in Siberia for the production of aviation fuel from coal. The plants at Leuna, Böhlen, Zeitz and Brüx continued with coal and coal tar hydrogenation into the early 1960s, but after modification they are also refining natural petroleum. Thus, today possibly four of Germany's twelve World War II coal hydrogenation plants have continued to hydrogenate coal. (32)

The Technical Oil Mission and Similar Investigative Groups

During the years Germany was developing its synthetic petroleum industry, the Roosevelt administration recognized the real possibility of American involvement in a second world war. As early as 1941, Roosevelt expressed concern over American petroleum reserves, and he established at that time the Office of Petroleum Coordinator for National Defense, to gather information on petroleum products and to make recommendations to insure an adequate petroleum supply. (By June 1941, private industry joined the program, lending technical advice to the government.) An executive order on December 2, 1942 created the Petroleum Administration for War (PAW) within the Office of Petroleum Coordinator, and from PAW originated the idea of sending technical experts to Germany to collect information on Germany's synthetic petroleum industry. In the summer of 1943 Harold Ickes, PAW administrator, made a formal recommendation to the Joint Chiefs of Staff to establish a Technical Oil Mission to Germany. (33)

The PAW expected private industry and the United States Bureau of Mines to capitalize on Germany's considerable experience in synthetic petroleum production. With the $30 million that Congress authorized in 1944 for a five-year synthetic petroleum research and development program, the Bureau planned to resume synthetic petroleum research where Germany had stopped. (34)

By December 11, 1944, advisors from industry had selected twenty-six petroleum scientists and engineers from private companies (Standard, Gulf, Phillips) and the Bureau of Mines for the Technical Oil Mission (TOM) to Germany. William C. Schroeder of the Bureau of Mines directed the TOM in its search for documents. (35) The first TOM members arrived in Germany in late February 1945 and entered the German plants as soon as possible after the plants fell into Allied hands. Although inspection of the plants was useful, the most valuable information the TOM obtained came from plant production documents, research documents, and interviews with German scientists and engineers such as Matthias Pier and Ernst Donath. (36) The amount of captured documents utterly amazed the TOM. When they entered the I.G. Farben building in Frankfurt, papers were knee-deep on the floors of the gigantic building and waist-deep in the stairwells. TOM collected the technical documents and sent them to London by the ton where members of the Combined Intelligence Objectives Subcommittee (CIOS) began translating and abstracting them.

As the flow inundated the CIOS offices, its staff could only microfilm the documents at random, producing 141 reels. Later the CIOS moved the entire program to the United States where it produced another 164 microfilm reels. Many documents that arrived in the United States found their way to loosely defined depositories, which often were simply the first convenient locations with available storage space. (37)

CIOS and two other intelligence gathering teams, BIOS (British Intelligence Objectives Subcommittee) and FIAT (Field Intelligence Agency, Technical) issued more than 1400 reports on the German synthetic petroleum plants. They also released the interviews of German scientists and engineers, written at the plant sites or shortly after in Paris and London. Many of these reports and interviews are on the TOM reels. (38)

Synthetic Petroleum Production in the United States

By the end of 1945, the Technical Oil Mission had largely disbanded and most of its members returned to their positions in private industry or with the Bureau of Mines Synthetic Liquid Fuels Division. The Bureau had done limited research on coal hydrogenation since 1924, but because of the abundance of natural petroleum in the United States and the high cost of Bergius's synthetic process it never actively encouraged development. Arno C. Fieldner (1881-1966), the Bureau's chief chemist in Washington, pointed out in 1928 that synthetic petroleum obtained from coal hydrogenation probably cost 40 to 50 cents a gallon and showed its "utter impossibility of approaching competition with petroleum gasoline costing 7 to 9 cents at the refinery, the prevailing present-day costs in the United States." (39) Fieldner found it "highly wasteful to consume a large portion of our present supply of petroleum for ordinary heating and stationary power generation, where coal would answer as well, and then find it necessary to replace this oil in a thermally expensive manner from coal. A forward-looking national fuel policy would seek to delay the day of making liquid fuel from coal as long as possible by reserving the higher value fuels of natural gas and petroleum for those uses which cannot be met so efficiently by the direct combustion of coal." (40)

Fieldner's attitude no doubt influenced the Bureau's position on high-pressure coal hydrogenation; but I.G. Farben's successful development of the Bergius process in the late 1930s convinced the Bureau that it should at least evaluate the German work on synthetic fuels and at the same time obtain data on the suitability of hydrogenating American coals. The Bureau believed that the results of its program would be valuable at some future time when the United States might have to supplement its natural and non-renewable petroleum reserves. Thus, in 1936 the Bureau constructed at its Central Experimental Station in Pittsburgh a small continuous unit which hydrogenated 100 pounds of coal in 24 hours. (41) By 1944, it built and operated two small pilot plants at Pittsburgh and Bruceton, Pennsylvania and was now ready to begin semicommercial development. (42) Funds for the semicommerical development came from Congress's $30 million authorization of 1944. The 78th Congress passed Public Law 290 which directed the Department of the Interior, Bureau of Mines, to carry out the production of synthetic liquid fuels from coal and other substances. Amendments in 1948 and in 1950 extended the program. (43)

With the end of World War II in August 1945, the United States War Department had available at Louisiana, Missouri a high-pressure synthetic ammonia plant formerly operated by the Hercules Powder Company. The Bureau of Mines acquired this plant on February 1, 1946 for the purpose of converting it to a semicommerical high-pressure coal hydrogenation plant. Construction started in May 1947 and ended two years later in 1949. The following year, the Bureau built a second synthetic fuel plant to test the Fischer-Tropsch method. Both plants were operational in 1951 and had the capacity to produce 200-400 barrels of synthetic petroleum per day. (41) In the research and development program at Louisiana, the Bureau utilized its almost thirty years of experience with synthetic petroleum production and the information contained in the captured German documents. The Defense Department also quietly brought seven German scientists to assist in the program. (45)

The research at Louisiana, Missouri demonstrated clearly the feasibility of hydrogenating American coals. Economically, however, coal-derived petroleum could not compete with natural petroleum. The manufacturing cost of synthetic gasoline, according to the Bureau in 1952-53, was 19.1 cents per gallon or almost double the 10.6 cents cost of regular grade gasoline at petroleum refineries. Other cost studies on synthetic gasoline questioned the accuracy of the Bureau's figures. Ebasco Services, a cost consulting firm in New York City, estimated 21.8 cents per gallon, while the National Petroleum Council arrived at 34.8 cents. The great variance in cost among the three studies resulted from widely different estimates of profits earned from the sales of the process's chemical by-products. The Bureau's program succeeded technologically, but the economics of coal hydrogenation precluded its continuation in a time of abundant and cheap natural petroluem. (46) In 1954 the Bureau terminated its program at Louisiana, Missouri and dismantled the coal liquefaction and Fischer-Tropsch plants.

Recent Developments in Coal Hydrogenation

During the past thirty years, private industry and government laboratories have continued research on coal hydrogenation. All of them used to some extent the older German technology. One of the first private coal hydrogenation research programs took place in the years 1952-56 when Union Carbide operated a 300-ton-per-day pilot plant at Institute, West Virginia. The $11 million high-pressure plant (P = 200-400 atm) incorporated the latest advances in coal conversion technology (such as accurate temperature control) and enabled Carbide's scientists to decrease reaction time from 45 to 3 minutes. Unlike the German hydrogenation plants, the Carbide plant produced primarily chemicals such as benzene, which was in short supply, phenols for use in plastics, pharmaceuticals derived from quinoline (nicotinic acid), picoline for tuberculosis drugs, and new rocket fuels. Carbide's goal was to develop new products from coal and thus permit a gradual return to coal as a primary chemical feedstock rather than natural gas. Carbide later planned to build a full-size, 1000 to 4000 tons-per-day plant to hydrogenate West Virginia's coal deposits but for economic reasons it closed the Institute plant in 1956. (47)

Nearly ten years later, in 1962, the Office of Coal Research (OCR), a branch of the United States Department of the Interior, awarded Willard Bull, a chemist with Spencer Chemical Company, a contract to continue the coal hydrogenation research that he had begun in the early 1950s. One of the government's reasons for renewed interest in coal-derived petroleum was that by this time the United States had gone from a net exporter of petroleum to a net importer. Spencer carried out small-scale research in its laboratory in Merriam, Kansas, converting up to 50 pounds of coal daily to a liquid product that solidified upon reducing the reaction temperature and pressure. Spencer called its coal hydrogenation process Solvent Refined Coal (SRC), for essentially the process consisted of mixing finely-crushed dried coal with a heavy oil solvent (1 part coal:2 parts oil) and then adding hydrogen gas at high pressure (69-137 atm) and temperature (450°C) to the coal slurry producing a black, clean-burning glassy solid boiler fuel. During the early research Spencer's chemists found that in lowering the ash content of the solid product to 0.5 weight percent they had removed most of the sulfur originally present in the coal. Thus they not only minimized the ash disposal problem at coal-burning power plants but had reduced substantially the primary contributor to air pollution. (48) By 1966, because of favorable technical and economic analyses and a continually deteriorating balance between petroleum consumption and domestic production, Spencer decided to expand its program. (49)

Gulf Oil Corporation had now acquired the Spencer Company, and in 1968 with support from a second OCR contract for $88 million, Gulf began design and construction of a 50-ton-per-day pilot plant at Fort Lewis, Washington, south of Tacoma. Gulf completed construction of the Fort Lewis plant in September 1974. The SRC pilot plant successfully hydrogenated high-sulfur bituminous coals from the eastern United States and bituminous coals from Pennsylvania, Kentucky, West Virginia, and Ohio. Its solid product, called SRC-I, had a heating value of 16,000 BTU/pound, which was 33 percent greater than that of the unhydrogenated coal. The process converted about 65 percent of the coal to the clean-burning solid, 15 percent to distillable liquids and the rest to ash and gases. (50)

Gulf's ongoing laboratory work early in the 1970s showed that modifying the SRC-I process to operate at a higher pressure (200 atm) and a slightly lower temperature (300-400°C) resulted in a greater hydrogen addition and gave a liquid rather than a solid product. The liquid resembled the petroleum boiler fuels burned in oil-fired electric power plants. Gulf found that their new, improved version of the SRC process, SRC-II, also produced the heavy oil solvent used to dissolve the finely-crushed dried coal, and eliminated the need to add constantly external solvents. (51)

The Fort Lewis plant began intermittent SRC-II production in 1974 converting daily about 30 tons of high-sulfur eastern bituminous coals chiefly to 70 barrels of home heating oil and fuels for use in public utility power plants. Indeed, in an experiment of August 1978, Consolidated Edison's 74th Street Station in New York City burned 4,500 barrels of SRC-II fuel oil from the Fort Lewis pilot plant to generate electrical energy. The coal-derived fuel oil performed as well technically and environmentally as petroleum-derived fuel oil. (52)

Because the Fort Lewis pilot plant was a success, the Department of Energy (DOE) decided to take the next step toward developing an American synthetic petroleum industry. In July 1978, it awarded Gulf a contract to design, construct, and operate a high-pressure coal hydrogenation demonstration plant. At the same time, in the fall of 1978, DOE agreed to share the hydrogenation technology developed in the demonstration plant with Japan and West Germany, two countries almost totally dependent on imported petroleum for their liquid fuels.

DOE investigated more than twenty coal producing areas in the United States before selecting a 2,600-acre site north of

Morgantown, West Virginia, on which to construct a plant processing 6,000 tons of coal daily thus producing about 18,000 barrels of synthetic petroleum each day. The plant's size though 120 times larger than Gulf's pilot plant was still one-fifth the estimated size of a future commercial production plant, hydrogenating 30,000 tons of coal per day. Gulf scheduled major construction to begin in the spring of 1982 with operation set for 1985. The plant's projected cost, $1.4 billion, certainly a major investment, but a figure that represented only the cost of six days' oil imports at the low price of $30 per barrel. (53)

Gulf intended to produce chiefly distillate fuel oil in the West Virginia plant. This is a low sulfur (less than 0.3 percent) nonpolluting fuel for the production of electrical power and steam in the eastern United States, where utilities and industry presently use natural petroleum fuel oil. Gulf claimed that a larger commercial-size plant processing 30,000 tons of coal daily would yield 60,000 barrels of distillate fuel oil or enough to meet the electrical demands of a city with one million inhabitants.

Naphtha was the second major product scheduled for production in the SRC-II demonstration plant. This lighter, low-boiling range hydrocarbon mixture (30-200°C), when upgraded, produces a high octane unleaded gasoline. A commercial-size plant's naphtha production is about 20,000 barrels (800,000 gallons) of gasoline daily, an amount consumed in a city of 250,000 population. (54)

DOE and Gulf planned a two-year demonstration period and if the operation was technically and economically successful, Gulf had the option to purchase the plant and proceed with commercial development. But in early 1981, one year before the start of constuction on the West Virginia plant, DOE decided not to continue funding the project. DOE's decision was a direct result of the government's new energy policy which calls for private industry to play a larger role in the commercial development of synthetic fuels. Upon learning of DOE's decision, the German, Japanese and American sponsors of the West Virginia plant met in Bonn, Germany to discuss the fate of the proposed demonstration plant. Faced with the loss of DOE support they decided to terminate the entire project. (55)

Two other coal hydrogenation projects started operation in the early 1960s. Unlike the cancelled West Virginia plant, they are active today. Both were larger than Gulf's initial project and also seemed to reflect an awareness of the United States's increasing dependence on imported natural petroleum. The first, the H-Coal Process, was a joint project initially involving the government and Hydrocarbon Research Inc., of Trenton, New Jersey. It began in 1963 in Hydrocarbon Research's laboratory with a 25-pound-per-day, bench-scale unit and then with a larger process development unit that hydrogenated three tons of coal daily. (56) By the early 1970s (1971-72), Ashland Synthetic Fuels, a subsidiary of Ashland Oil Inc., Ashland, Kentucky, and Badger Plants, Inc., Cambridge, Massachusetts, joined the project. (57) In the H-Coal process, crushed, dried coal is mixed with a heavy oil derived from the process to produce a slurry and pumped to a pressure of about 200 atm. After adding compressed hydrogen gas to the slurry and heating to 345-425°C (650-700°F), the slurry passes into an ebullated-bed reactor containing a metallic catalyst (cobalt-molybdenum alumina extradate) where liquefaction occurs. All hydrogenation processes require large amounts of hydrogen gas. The H-Coal Process consumes 14,000-20,000 standard cubic feet of hydrogen for each ton of coal hydrogenated and converts 90 percent of moisture- and ash-free coal to synthetic petroleum. Each ton of coal produces three barrels of synthetic liquid. (58)

The Office of Coal Research, now a division of DOE, originally provided funds for Hydrocarbon Research's early work. By the mid-1970s, it decided that the H-Coal Process had progressed to the pilot plant stage. In late 1976, Hydrocarbon Research and its partners began construction on a 200 to 600 ton-per-day plant in Catlettsburg, Kentucky, on a 50-acre site adjacent to Ashland Oil's refinery. Coal to liquid conversion started in May 1980. The plant reached its 600-ton capacity in 1981, converting coals from all over the country to liquid fuels for transportation, home and industry. DOE originally provided $143 million of the project's $179 million anticipated cost, but cost overruns raised the final price to $296 million. The final step in the H-Coal Project, if economic evaluations are favorable after a two-year test period, will be the construction of a $1.4 billion commerical plant with a synthetic petroleum capacity of 50,000 barrels per day. (59)

The remaining coal hydrogenation pilot plant in the United States is the recently completed Exxon plant in Baytown, Texas. (60) It represents the culmination of pilot plant studies which began in 1966 with units of one-half and one ton-per-day capacities. The Baytown plant, located on a 55-acre site adjacent to the Exxon refinery, is another joint industry/government operation with 50 percent of the $340 million cost coming from DOE. Exxon is the largest industrial participant (23.2 percent); two of the other participants, the Japan Coal Liquefaction Development Company (8.3 percent) and Ruhrkohle A. G. of West Germany (2.7 percent) represent interests outside of the United States. The plant which began operation on June 24, 1980, after a six-year design and construction period, will convert 250 tons of coal per day into approximately 600 barrels of synthetic petroleum using a third coal conversion method called the Exxon Donor Solvent Process (EDS). (61)

Like the other coal hydrogenation processes, SRC-II and H-Coal, the Donor Solvent Process requires high pressures and temperatures (T = 425-470°C [800-880°F], P = 130-135 atm [1500-2000 psi]). It differs from the other processes because the solvent oil added to the crushed dried coal to produce a paste or slurry is a previously catalytically hydrogenated recycle liquid such as tetralin, $C_{10}H_{12}$. Thus, the liquefaction of coal results not only from the addition of hydrogen gas to the slurry but also because the solvent oil donates some of its hydrogen atoms to the coal. According to Exxon, use of a hydrogenated rather than an unhydrogenated solvent greatly improves the liquefaction process. The metallic catalyst (nickel-molybdenum), since it does not contact coal minerals or high-boiling liquids, has a longer, more active lifetime. (62)

The Baytown plant is self-sufficient, providing its own fuel and the hydrogen gas required for liquefaction. Gasifying the coke obtained in the hydrogenation process supplies the fuel, reforming the light gaseous products, CH_4 and C_2H_6, by reacting them with steam, produces hydrogen gas. The EDS process has about a 71 percent yield of dry coal to synthetic petroleum, converting one ton of bituminous coals, sub-bituminous coals or lignites to 2.7-3.1 barrels of low-sulfur, gasoline-size molecules

$$C_4\text{-}C_{15},\ BP = 200\text{-}300°C$$

and lubricating oil

$$C_{15}\text{-}C_{25},\ BP = 300\text{-}400°C$$

The pilot plant will operate until 1982 and if successful, Exxon will begin commercial-size development in 1990. (63)

Conclusion

Since 1958 the United States has been a net importer of petroleum. Though domestic consumption continued to increase in the 1960s, only in 1964-1965 did it begin to increase more rapidly, going from approximately four billion barrels per year to almost five and one-half billion in 1970. The United States's petroleum production increased in this period but more slowly than demand. In 1970 domestic production reached its peak and began to decline. Within two years, in 1972, the United States experienced its first trade deficit since 1935 and only the second deficit since 1888. (64)

The mid-1960s also marked the beginning of basic industrial research on coal liquefaction, perhaps in anticipation of future petroleum shortages. But the drive to develop an American synthetic fuel industry did not occur until after the Arab oil embargo during the winter of 1973-74. Today, coal liquefaction seems on the way to becoming a successful technological and economical alternate energy source. All three conversion processes, according to cost estimates made in mid-1979, produce a synthetic petroleum for less than $30 per barrel. (65) Other energy sources, such as solar and nuclear, can provide thermal energy, mechanical energy or produce electricity but none can supply liquid fuels and lubricants.

Of course, coal liquefaction can never be a permanent solution to the United States's energy problems, for coal, though presently plentiful, is nevertheless a nonrenewable resource with a finite future. The U.S. Geological Survey in 1976 identified American coal resources at approximately 1,731 billion tons, with a demonstrated reserve base of 438 billion tons which existing technology can mine with a recovery of at least 50 percent. American coal reserves far exceed the combined known reserves of all other fossil fuels (oil, natural gas, oil shale and tar sands) and have an estimated lifetime of 200-300 years. (66)

There are also environmental and health factors to consider before developing a coal liquefaction industry. Are the increased amounts of carbon dioxide escaping into the atmosphere during liquefaction great enough to render the process a high-risk technology? Some scientists fear that a sufficient increase of atmospheric carbon dioxide will produce a greenhouse effect resulting in a warming of the earth's surface and a subsequent shift in the earth's agricultural zones, variance in rainfall, changes in marine production, and a rise in sea levels. Sulfur dioxide emission will not be a problem because the liquefaction process removes most of the sulfur present in the coal. (67)

In addition to the greenhouse effect, what will happen to a geographic area after having mined thousands of tons of coal for use in synthetic petroleum production? Is it possible to restore vegetation to a mined-out area? Coal liquefaction will need large quantities of water for power plant cooling towers, for generating electricity and for manufacturing the hydrogen gas essential for liquefaction. A 6,000-ton-per-day demonstration plant requires 4,000 gallons per minute (1.1 cubic feet per second) or about four gallons of water for every gallon of synthetic petroleum. Will a limited water supply in certain coal-bearing regions of the West hamper the development of a coal hydrogenation industry? Alternatively, can reclamation processes eliminate possible water pollution caused by waste water discharge into nearby streams? (68)

Finally, there is danger of carcinogenic hydrocarbon emission. During the 1950s, workers at Union Carbide's coal liquefaction plant at Institute, West Virginia developed skin cancer. A follow-up study completed during the late 1970s showed, however, that none of the workers developed systemic (internal) cancers twenty years later. Thus, exposure to coal liquefaction chemicals appeared to cause no increase of death due to systemic cancers. Indeed, many biomedical experts believe that because present worker protection procedures are so superior to the primitive techniques used at the Institute plant, the incidence of skin cancer will decrease. (69)

DOE and private industry recognize today the difficulties associated with developing a synthetic petroleum industry. DOE has withdrawn funding for the proposed Morgantown, West Virginia plant, but through its Synthetic Fuels Corporation, DOE intends to provide substantial support for industry in the form of loan and price guarantees. (70) A major reason for this decision is that the United States contains the greatest amount of the world's known recoverable coal reserves, 31 percent or 250 billion tons. (71) The government's long-range goal, according to the Energy Security Act of July 1980, was to build twenty commercial plants, each costing about $1.5-$2.0 billion which by 1992 would have produced two million barrels per day. (72) The present plan calls for production of 500,000 barrels by that date. (73) This figure represents only a small fraction of the United States's daily petroleum consumption but clearly contributes toward lessening the uncertainty of our energy future.

Acknowledgments

The National Science Foundation, Grant No. SES 792648, and the Center for Energy and Mineral Resources, Texas A&M University have provided funds for this research.

Literature Cited

1. Duckert, Joseph M. "A Short Energy History of the United States"; The Edison Electric Institute: Washington, DC, 1980.
2. Fieldner, Arno C. Industrial and Engineering Chemistry 1926, 18, 1009; Wigginton, R. Fuel 1925, 4, 95.
3. Bergius, Friedrich. Industrial and Engineering Chemistry, news ed., December 10, 1926, 4, 9-19; Journal of the Institute of Fuel 1934, 8, 74-79; Proceedings of the World Petroleum Congress, Volume II, 1933, pp 282-89; "Nobel Lectures, Chemistry, 1922-1941"; Elsevier Publishing Company: New York, 1966; pp 244-76.
4. Taylor, F. Sherwood. "A History of Industrial Chemistry"; Arno Press: New York, 1972; pp 428-34.
5. Bergius, Friedrich. J. Soc. Chem. Ind. 1913, 32, 462-67; "Die Anwendung hoher Drucke bei chemische Vorgangen und eine Nachbildung des Entstehungsprozesses der Steinkohle"; Wilhelm Knapp: Halle, 1913; German Patents 304,348 September 17, 1919 (filed May 6, 1913); 301,231 August 9, 1913. British Patents 18,232 August 1, 1914; 5021 March 31, 1915.
6. Bergius. Proceedings of the World Petroleum Congress, pp 282-89. See also the patents listed in note 5.
7. German Patent 301,231, August 9, 1913.
8. Bergius, Friedrich. Zeitschrift für angewandte Chemie 1921, 34, 341-47. Because of the outbreak of World War I and perhaps because of the delay in his receiving patents on coal hydrogenation, Bergius published nothing on his process until 1921. This is the first article along with American patents 1,342,790, June 8, 1929 (filed April 18, 1916) and 1,291,664, September 27, 1921 (filed April 18, 1916) in which Bergius discussed his continuous process.
9. Ritter, Gerhard. "The Schlieffen Plan: Critique of a Myth"; Praeger Press: New York, 1958; Schlieffen, Alfried Graf von. "Gesammelte Werke"; 2 vols., E. S. Mittler: Berlin, 1913.
10. Bosch, Karl. Chemisches Zentralblatt 1913, 2, 195; German Patents 249,447, July 23, 1912 and 254,437, December 3, 1912; Chemical Age 1921, 29, 451-45; The Industrial Chemist 1934, 10, 90-94; "Nobel Lectures, Chemistry, 1922-1941"; Elsevier Publishing Company: New York, 1966; pp 195-235; Mittasch, Alwin.

"Geschichte der Ammoniaksynthese"; Verlag Chemie: Weinheim, 1951; pp 93-116; Mittasch, Alwin and Frankenburger, W. Zeit. für Elektrochemie und angewandte physikalische Chemie 1929, 35, pp 920-27; Haber, Ludwig. "The Chemical Industry 1900-1930"; Clarendon Press: Oxford, 1971, pp 93-95; Farber, Eduard. Chymia 1966, 11, pp 157-58.

11. The Iron and Coal Trades Review, December 3, 1927, 135, p 937.
12. Ibid.
13. The Chemical Age 1926, 15, p 150.
14. Bergius. Proceedings of the World Petroleum Congress, p 287.
15. Bergius. J. Soc. Chem. Ind. 1913, 32, p 463; German Patents 254,593, December 9, 1912 and 277,501, November 30, 1913; U.S. Patent, 1,059,818, April 22, 1913.
16. Bergius. Proceedings of the World Petroleum Congress, p 289.
17. Ibid.
18. "High-Pressure Hydrogenation at Ludwigshafen-Heidelberg," FIAT Final Report No. 1317, Volume I, published for Joint Intelligence Objectives Agency by Central Air Documents Office, Dayton, Ohio, 1951, pp 24-26.
19. Ibid., p 32.
20. "Report on the Petroleum and Synthetic Oil Industry of Germany"; The Ministry of Fuel and Power; His Majesty's Stationery Office: London, 1947, pp 46-69.
21. Ibid., pp 46-48.
22. Fischer, Franz and Tropsch, Hans. Brennstoff-Chemie 1923, 4, pp 192-97; Brennstoff-Chemie 1923, 4, pp 276-85; Fischer, Franz. Ind. and Eng. Chem. 1925, 17, pp 574-75; Fischer and Tropsch. Brennstoff-Chemie 1925, 6, pp 233-34; Brennstoff-Chemie 1926, 7, pp 97-104.
23. "Report on the Petroleum and Synthetic Oil Industry of Germany"; pp 1-2, p 50.
24. Prewar rate of exchange: RM = $.25. The exchange rate is from Statistical Abstract of the United States, 1939, Washington: United States Government Printing Office, 1939, p 208. To convert tons of petroleum to barrels of petroleum: 1 ton = 7.3 barrels, 1 barrel = 42 U.S. gallons.
25. 100 Pfenning = 1 RM.
26. U.S. Army Military Court, Nuremberg, Transcripts of Proceedings, Case VI (Farben) U.S. vs. Carl Krauch et al., 24 (in German), 5063, Testimony-Krauch; the Fuel agreement is Nuremberg Document NI-881; see also Birkenfeld, Wolfgang. Tradition 1963, 8, pp 107-108; Birkenfeld, "Der synthestische Triebstoff 1933-1945"; Musterschmidt-Verlag: Göttingen, 1964, pp 23-34.
27. Birkenfeld. "Der synthetische Triebstoff"; pp 77-137.
28. "Principles to Govern the Treatment of Germany in the Initial Control Period," No. 848, The Conference of Berlin (The Potsdam Conference), 1945, Foreign Relations of the United States, Diplomatic Papers, Volume II, United States Government Printing Office: Washington, DC, 1960, pp 750-53, see p 752.

29. "Multilateral German Industries," United States Treaties and Other International Agreements, Volume II, Part I, United States Government Printing Office: Washington, DC, 1951, pp 962-72, see p 963.
30. "Multilateral Incorporation of Germany into European Community of Nations," United States Treaties and Other International Agreements, Volume III, Part II, United States Government Printing Office: Washington, DC, 1954, pp 2714-22, see p 2716.
31. Birkenfeld. "Der synthetische Triebstoff"; pp 213-15.
32. Ibid.
33. Miller, Albert E. "The Story of the Technical Oil Mission." Reprint of the Twenty-fifth Annual Meeting of the American Petroleum Institute, Chicago, Illinois, November 14, 1945.
34. U.S. Statutes at Large, 78th Congress, 1944, Vol. 58, Part 1, pp 190-91.
35. Miller. "The Story of the Technical Oil Mission," pp 1-8.
36. Schroeder, W. C. The Oil and Gas Journal 1945, 44, pp 112-116.
37. Miller. "The Story of the Technical Oil Mission," p 13.
38. Storch, H. H.; Golumbic, Norma; Anderson, Robert. "The Fischer-Tropsch and Related Syntheses"; John Wiley and Sons: New York, 1951; p i.
39. Fieldner, Arno C. Fuel 1928, 7, p 497.
40. Fieldner, Arno C. Ind. and Eng. Chem. 1935, 27, p 988.
41. Hirst, L.L.; Haw, C.O.; Sprunk, G.C.; Golden, P.L.; Pinkel, I.I.; Boyer, R.L.; Schaeffer, J.R.; Kallenberger, R.H.; Hamilton, H.A.; Storch, H.H. Ind. and Eng. Chem. 1939, 31, pp 869-77; Storch, H.H.; Hirst, L.L.; Fischer, C.H.; Sprunk, G.C. "Hydrogenation and Liquefaction of Coal," Bureau of Mines Technical Paper 622, 1941; Storch, H.H.; Hirst, L.L.; Fisher, C.H.; Work H.K.; Wagner, F.W. Ind. and Eng. Chem. 1941, 33, pp 264-74; Storch, H.H.; Hirst, L.L.; Golden, P.L.; Pinkel, I.I.; Boyer, R.L.; Schaeffer, J.R.; Kallenberger, R.H. Ind. and Eng. Chem. 1937, 29, pp 1377-80.
42. Markovits, J.A.; Braun, K.C.; Donovan J.T.; Sandaker, J.H. "Special Equipment in the Coal-Hydrogenation Demonstration Plant," Bureau of Mines Rept. Inv. 4584, 1950.
43. See note 34. Also, Wu, W.R.K.; Storch, H.H. "Hydrogenation of Coal and Tar," Bureau of Mines Bulletin 633, 1968.
44. Markovits, Braun, Donovan, Sandaker. "Special Equipment in the Coal-Hydrogenation Demonstration Plant;" Grunder, L.J. "Report on the Hydrogenation Demonstration Plant Division, Louisiana, Missouri, unpublished report of March 13, 1946. Reports on the Bureau's Louisiana, Mo. plant are part of the German Document Retrieval Projects Collection, Center for Energy and Mineral Resources, Texas A&M University, College Station, Texas, 77843; Kastens, Merritt L.; Hirst, L.L.; Chaffee, C.C. Ind. and Eng. Chem. 1949, 41, pp 870-85.

45. A total of sixteen German scientists worked at the Louisiana, Missouri plant, five in coal hydrogenation, two in Fischer-Tropsch, eight in oil shale, one in acetylene production. See J.L. Grunder's Report, note 44.
46. Hirst, L.L.; Markovits, J.A.; Skinner, L.C.; Dougherty, R.W.; Donath, E.E. "Estimated Plant and Operating Costs for Producing Gasoline by Coal Hydrogenation," Bureau of Mines Rept. Inv. 4564, 1949; Wu and Storch, "Hydrogenation of Coal and Tar," p 6.
47. Chemical and Engineering News 1952, 30, pp 154-59; Chemical and Engineering News 1956, 34, p 4009; Lessing, Lawrence P. Scientific American 1955, 193, pp 38-67.
48. Gulf Oil Corporation, Information Brochure, "Solvent Refined Coal-II: Turning Coal into Clean Energy," pp 4-5; Assessment of Technology for the Liquefaction of Coal, The National Research Council, Washington, 1977, pp 47-50.
49. Gulf Oil Corporation, Information Brochure, "Solvent Refined Coal-II: Background and Schedule."
50. See note 48.
51. Gulf Oil Corporation, Information Brochure, "Solvent Refined Coal-II: Turning Coal into Clean Energy," pp 5-7; Information Brochure, "Solvent Refined Coal-II: The Process."
52. Gulf Oil Corporation, Information Brochure, "Solvent Refined Coal-II: Background and Schedule."
53. Ibid.; Oil and Gas Journal 1980, 78, p 33.
54. Gulf Oil Corporation, Information Brochure, "Solvent Refined Coal-II: Products."
55. Energy Insider 1981, 4, p 2.
56. Schuman, Seymour C.; Wolk, Ronald H.; Chervenak, Michael C. U.S. Patent 3,183,180, May 11, 1965; Alpert, S.B.; Johanson, E.S.; Schuman, S.C. Hydrocarbon Processing 1964, 43, pp 193-96.
57. Ashland Synthetic Fuels, Inc., Information Brochure, H-Coal, A Future Energy Source, November 1979; "H-Coal Special Issue," Department of Energy, Energy Countdown, November 1979; Comoli, Alfred G.; Battista, Carmine A.; Johanson, Edwin S.; Laird, Carol P., "Development and Demonstration of the H-Coal Process," paper presented at the Division of Petroleum Chemistry, American Chemical Society Meeting, Miami, September 11-15, 1978.
58. Schuman, Seymour, C.; Wolk, Ronald H.; Chervenak, Michael C. "Hydrogenation of Coal," U.S. Patent 3,321,393, May 23, 1967; "The H-Coal Process," January 9, 1981, paper submitted by Hydrocarbon Research, Inc. to the Research and Development Council of New Jersey for the Thomas A. Edison Award; Comoli, Alfred G.; Ganguli P.; Merdinger, Michael, "H-Coal Processing of Kentucky No. 11 Coal and 1980 Status of H-Coal," 15th Intersociety Energy Conversion Engineering Conference, Seattle, Washington, August 18-22, 1980, pp 1818-23; Eccles, Richard M.; DeVaux, George R., "Current Status of H-Coal Commercialization,"

presented at the Syncrude Fuels Conference, San Francisco, California, October 13-16, 1980. Assessment of Technology for the Liquefaction of Coal, The National Research Council, Washington, 1977, p 66.

59. H-Coal, A Future Energy Source, "H-Coal Special Issue," Energy Countdown; Eccles and DeVaux, "Current Status of H-Coal Commercialization;" "H-Coal Plant Sets February Restart Using Western Sub-Bituminous Coal," Energy Insider, February 1982, 5, p 1.
60. The Fluor Corporation recently reactivated and rebuilt the DOE test plant at Cresap, West Virginia. It conducted a coal liquefaction test project for DOE using Exxon's Donor Solvent Process. Oil and Gas Journal 1980, 78, p 71.
61. "Exxon Coal Liquefaction Plant Prepared for Startup," Coal R&D 3 (February 15, 1980), pp 1-2; Porter, Robert. "April Startup set for Synfuel Plant in Texas," Energy Insider 3 (February 18, 1980), pp 1, 5; 3 (July 7, 1980), p 1; Furlong, L.E.; Effron, E.; Vernon, L.W.; Wilson, E.L. Chemical Engineering Process 1976, 72, pp 69-75; Chemical and Engineering News 1980, 58, pp 7-8; Assessment of Technology for the Liquefaction of Coal, The National Research Council, Washington, 1977, pp 55-57.
62. Taunton, John W.; Trachte, K.L.; Williams, R.D. Fuel 1981, 60, pp 788-94; Lewis, Debra, "Information an Important Product at DOE's Synfuel Plant in Baytown, Texas," Energy Insider, September 7, 1981, 4, pp 4-5; Vick, G.K.; Epperly, W.R. Science 1982, 217, pp 311-16.
63. Porter, Robert. "Second Synfuels Plant Now Operation in Texas." Energy Insider, July 7, 1980, 3, p 1.
64. Duckert. "A Short Energy History of the United States"; p 63.
65. Rogers, K.A.; Hill, R.F. Coal Conversion Comparisons. The Engineering Societies Commission on Energy, Inc.: Washington, July 1979, pp 49-60.
66. "Coal: Energy Source Past and Future," Energy Insider, December 8, 1980, 3, p 2.
67. Scriven, R.A. Journal of the Institute of Energy 1980, 53, pp 15-16; Madden Roland A.; Ramanthan, V. Science 1980, 209, pp 763-68; MacDonald, Gordon J. The Physics Teacher 1979, 17, pp 367-73.
68. Chemical Week 1980, 128, pp 25-27. Rogers and Hill, Coal Conversion Comparisons, pp 69-71; Assessment of Technology for the Liquefaction of Coal, pp 111-15; Gulf Oil Corporation Information Brochure, "Solvent Refined Coal-II Environmental Issues and Protection Measures."
69. Palmer, Alan. Journal of Organizational Medicine 1979, 21, pp 41-44; Assessment of the Technology for the Liquefaction of Coal, pp 115-119; Gulf Oil Corporation Information Brochure, "Solvent Refined Coal-II: Health Issues and Protection Measures."

70. Threlkeld, James. "Free Market Emphasis Will Reduce Need for Government Regulations," Energy Insider, August 3, 1981, 4, pp 3-6.
71. World Energy Conference Survey of Energy Resources. Compiled by Harold E. Goeller. New York: United States National Committee of the World Energy Conference. 1974.
72. Burke, Donald P. Chemical Week 1980, 128, pp 18-20.
73. Science 1981, 213, p 521.

RECEIVED January 7, 1983

3

From Unit Operations to Unit Processes

Ambiguities of Success and Failure in Chemical Engineering

JEAN-CLAUDE GUÉDON

Université de Montréal, Institut d'Histoire et de Sociopolitique des Sciences, Montréal, Canada

> "Chemistry views the chemical reactions; chemical engineering views the pocketbook reactions."
> Charles M. Stine (1928).

I. Background

Despite its present stage of relative infancy, the history of chemical engineering already harbors a number of important themes. One of them is that chemical engineering, as this word is now used in the United States and a number of other countries, refers to a disciplinary and professional structure which saw its original cognitive foundations based on a new notion - that of unit operations. This point has been developed elsewhere (1) but it will be useful to summarize some of the main elements of this story to provide a suitable backdrop for the present paper.

It should be recalled that the American Institute of Chemical Engineers (AIChE) was organized mainly to respond to the professional needs of a number of applied and industrial chemists who felt that they were not sufficiently heard by the American Chemical Society (ACS). But professional aspirations do not amount to a profession; a sense of professional identity is also necessary and this identity, in time, cannot exist without some clear definition of an area of competence where both know-how and knowledge are present. In effect, the would-be profession needs to rest on a well defined discipline to achieve some degree of maturity. In the case of AIChE, its founding fathers quickly focused their attention on educational matters and, in particular, they sought to design a suitable curriculum for future chemical engineers. In doing so they ultimately - albeit in a somewhat sleep-walking manner - succeeded in elaborating a discipline for themselves.

In 1915, a consulting engineer from Cambridge, MA, A. D. Little, submitted a report to the Corporation of the Massashusetts Institute of Technology in which he suggested a new approach to the teaching of chemical engineering (2). This approach was based on the notion of unit operations and it structured his new curricular design - The School of Chemical Engineering Practice.

0097-6156/83/0228-0043 $06.00/0

The importance of unit operations was immediately recognized by many practitioners of chemical engineering and the quick acceptance of A. D. Little's scheme can be accounted for by the several advantages it offered to the fledgling profession. First of all, it provided a clear distinction between chemical engineering and either chemistry or mechanical engineering. As a result, it reduced the pressure of competing domains of the young profession by clearly demarcating the cognitive domain of chemical engineering from that of chemistry or chemical engineering. A. D. Little's proposal was also attractive because it helped engineers to respond to the quickly growing variety of demands emerging from a chemical industry which was literally booming during and immediately after the first world war. Finally, it also allowed for a useful division of labor between universities and industries at the level of research. While universities could work with theoretical and general models of unit operations, industries could take these models and apply them directly to the working conditions of plant operations.

The success of unit operations can be related, in the end, to its ability to serve the various interests of a professional community, an academic community, and a complex industrial sector. Not surprisingly, therefore, unit operations become the new structural model of chemical engineering in the years following the first world conflict. But this success was not due solely to the happy convergence of a variety of needs expressed by the three social partners identified above. Actually, AIChE insured the victory of unit operations over other schemes by waging a vigorous campaign to impose it on universities. The word "impose" is not too strong if the tactics used by AIChE are examined with some care. These tactics were new among engineering, but not among older professions such as lawyers and doctors. They rested on one key device- accreditation.

Chemical engineers were the first among engineers to import accreditation as a means to put pressure on reluctant university departments. As early as 1921, A. D. Little, once again, published the results of a survey of chemical engineering education, and he concluded that report by stating that curricula should be quickly standardized to put an end to the anarchic situation revealed by the survey. The model to be emulated, of course, was his own as it had been realized at M.I.T. (3).

Instruments favoring the standardization of chemical engineering curricula quickly multipled after 1921. First of all, a textbook based on unit operations was needed. Principles of Chemical Engineering written by W. H. Walker, W. K. Lewis, and W. H. McAdams appeared in 1923. Two years later, the first list of accredited institutions was drawn up, and its impact must have been rather great since its main characteristic was strict selectivity. Only fourteen universities shared the honor of appearing in the select list of AIChE. This policy was not a temporary one. Ten years later, only ten more universities had earned the right to satisfy AIChE's stringent conditions. Of course, further research might show that many universities refused to play AIChE's game and disdained applying - a hypothesis partially supported by the fact that not a single Ivy League university was accredited before 1936. Accreditation stayed, however, and all universities knuckled under (4).

The unusual length of this introduction can be justified by the fact that it allows raising the question which will guide the remainder of this paper. In summary, it can be said that the disciplinary - professional structure of chemical engineering is made possible by the emergence and success of a new notion called unit operations. Both the emergence and success of unit operations have been demonstrated to result from a complex series of interactions between the industrial sector, the academic system, and professional organizations. What remains to be seen is the evolution of such a notion once it has been constructed. In other words, exactly as historians of science unravel the mode of evolution of certain objects such as concepts or theories, historians of technology may well have at their disposal kinds of objects which, although somewhat different from scientific concepts and theories, may nevertheless play an analogous role. The notion of unit operation lends itself to this exploration, and its extension to the notion of unit processes offers a favorable terrain to start exploring the historical characteristics of technical notions - notion being used here in contradistinction with concept and theory on the one hand, and model on the other (5,6).

II. Unit Processes: Chronology and Definition

Although the phrase "unit processes" has disappeared from the memory of many chemical engineers, it is not very difficult to retrace the main events dotting its uncertain career as R. N. Shreve (7) has recounted it several times. The term itself was coined as early as 1928 by P. H. Groggins, and only two years later Shreve himself was teaching courses organized around unit processes. The full institutionalization of unit processes took place between 1935 and 1937. It is marked by one important event in each of these years. In 1935, P. H. Groggins wrote the standard survey of the field corresponding to this new notion: Unit Processes in Organic Synthesis (8,9). The following year, AIChE modified its definition of chemical engineering to include unit processes next to the unit operations which, by then, had reached a quasi-canonical status in the discipline (10). Then, in 1937, "unit processes symposia" were inaugurated under the presidency of Shreve himself and went on until the early 1950s, on a yearly basis.

The best definition of unit processes was provided by R. N. Shreve in 1940. By then the notion had been tested and criticized in the field and, consequently, had reached something like a state of maturity.

> By 'unit processes' we mean the commercialization of a chemical reaction under such conditions as to be economically profitable. This naturally includes the machinery needed and the economics involved as well as the physical and chemical phases. But here we lay stress upon the chemical changes and upon the equipment and conditions necessary to affect these changes economically in distinction from the unit operations involving specifically the physical changes (11).

Unit processes, as is clear from the definition above, deals with the specifically chemical aspects of chemical engineering and leaves the physical and mechanical transformations of substances to unit operations. But this very distinction between the chemical and physical dimensions of chemical engineering brings some questions to mind. For one thing, it may sound paradoxical to state that before unit processes, chemical engineering had little if anything to do with chemical reactions. If this is true, what were the factors which led chemical engineers to focus attention on specifically chemical problems? Finally, a more profound question cannot be avoided: is the distinction between the chemical and physical aspects of chemical engineering a necessary consequence of the nature of this particular technical discipline, or is it the result of some temporary aspects of its general evolution?

III. The Physical and Chemical Dimensions of Chemical Engineering

With unit processes the emphasis was placed on chemical reactions. As a result, unit processes both differed from unit operations and complemented them. But does this mean that chemical engineering ignored chemical reactions before the advent of unit processes? Two contradictory answers to this question were given. The first answer is negative and tries to show that the chemical dimension had always been part of chemical engineering. In 1951, for example, T. H. Chilton was arguing that A. D. Little had not originally coined the phrase "unit operations"; but the phrase "unit action" and that this latter expression was meant to include both physical and chemical processes (12). Chilton may have been right, but A. D. Little's formulation may also reflect a certain indifference to the distinction between physical and chemical processes rather than a explicit will to include chemical processes. This indifference would account nicely for the various synonyms being used in the early 1920s. Chilton mentions "unit actions", but others were used as well, including "unit processes." For example, when Chemical and Metallurgical Engineering devoted a whole issue to unit operations in 1923, it called them "unit processes" (13).

The second answer to our question is more interesting. Chemical engineering not only had ignored chemical reactions, it should ignore them. This position was defended by Martin W. Ittner in reaction to a presentation of unit processes by D. B. Keyes in 1936. It was done in such clear terms that it deserves being quoted in full, as a kind of benchmark in this regard.

> In this matter of differentiating between physical and chemical operations, I wonder if any of you have thought of it in a way that I have thought of it: we do not, any of us, carry on any chemical operations. That is something that is beyond our control. All we can do is to carry on physical operations. We make a study of physical materials and we bring certain materials together in physical operations, and if we get them under certain conditions, a chemical reaction takes place. But we have no control over that at all. All we do is the physical operation, bringing together under other conditions, another chemical reaction takes place and the Lord himself carries on the chemical reaction. After the reaction has taken place we make a physical study of what has happened and, there again, we carry on physical operations (14).

Ittner's analysis displays the familiar contrast between control and knowledge which often demarcated the engineer's work from that of the scientist. But, more specifically, it also denies any role to unit processes defined as chemical processes. For Ittner, the unit process, once it is examined closely, can be resolved into two component parts, one physical and the other chemical. The former corresponds to the engineer's domain while the latter is under the control of nature's laws or, as Ittner says, the Lord. To be sure, the chemical engineer could not act if chemical laws did not exist, but insofar as his goal is control and not knowledge, he must be content with bringing substances together and then let nature run its course.

With regard to the chemical domain, the attitudes of chemical engineers were not as unambiguously favorable as might be expected at first (15). This is not to say that indifference or even hostility to chemistry were dominant features of chemical engineering, but the existence of these currents of thought is enough to show that the extension of unit operations into the realm of chemical reactions is not a step that can simply be expected. At any rate, it certainly does not necessarily stem from the essence of unit operations, so to speak.

IV. Responding to New Industrial Demands

In trying to understand why some chemical engineers sought to complete the notion of unit operation by a new notion patterned after this first one, but dealing specifically with chemical reactions, it is useful to turn to the wider industrial context in which these events were taking place. The picture of American chemical industries around and after the first world war is extremely complex, but only some aspects of it are relevant for the case at hand. Consequently, it will be easier to examine how the promoters of unit processes perceived this new industrial context. Three of them were particularly important in this regard. They are R. N. Shreve, P. H. Groggins, and J. B. Key

As early as October 1918, R. N. Shreve had organized (16) a symposium on synthetic dyes (17) with one goal in mind: he wanted to promote the creation of a new division within the American Chemical Society so as to coordinate technical and economic debates about a field which had amply demonstrated its crucial importance in the conflict then about to end (18). Shreve was acutely aware of the perils ahead for this young industrial sector, especially since peace would signal renewed and exacerbated competition among the great industrial nations of the world. The American dye industry needed some protection to face its older, more experienced rivals from Europe. In practice this meant high tariff barriers for a while and the promotion of industrial research. It must be recalled that the American dye industry was exceedingly small before 1914. About 300 or 400 different dyes were imported each year, and about 120 dyes were synthesized from imported intermediate compounds, most of which cam from Germany. To put it another way, in 1914, the American dye industry satisfied only 10% of the local needs and was entirely dependent on foreign suppliers for intermediate products (19).

When the first world war started, no one thought that it would last four years. As a result, the American dye industry started reacting to the new context by living off its stocks and by improvising. As for imports, they rapidly fell to nothing because of the naval blockade of German harbors. Such a state of unpreparedness could lead to only one result. By January 1915, the United States was limited to sixteen synthetic dyes, with regrettable esthetic consequences and, more importantly, regrettable consequences for those sectors of the textile industry sensitive to fast changing demands of fashion (20).

At that point the United States started to build an indigenous dye industry which, by the end of the war, could roughly produce anything that competing nations could produce. This industry had grown in something like a hot house atmosphere because it had had no rivals for three years. As a result, the industry was not yet competitive when peace came at last and took about a decade to become so. This qualifier should not demean the achievement of the American chemical industry between 1915 and 1918. It responded exceedingly well to the sudden and brutal demands of the war context and, within three years, delivered a whole variety of goods that it had been incapable of producing before. But this gigantic effort brought several consequences including a complete technical mutation. The American chemical industry used to be a heavy chemical industry based on the production of relatively few mineral compounds at the lowest possible cost. In the narrow compass of three years, the industry had added a capacity to produce many complex organic substances. To do so, however, it had to learn how to produce in entirely new ways, and to design and to use entirely new types of reactors. The tasks of the engineer had also been profoundly modified by these new demands (20,21,22).

American chemical engineers were not very well equipped to address such sudden problems in their industry. A. D. Little's formulation of unit operations, which appeared in 1915, could have but a limited impact on the growing organic industry because, in practice at least, it largely ignored purely chemical problems (23). In retrospect, this notion of unit operation appears as one which responded extremely well to the industrial demands of a rapidly disappearing past - namely those of a heavy chemical industry based on minerals and petroleum refining.

While R. N. Shreve seems to have been particularly sensitive to the explosive growth of the dye industry, P. H. Groggins' interests were limited to other facets of the new organic industry. He had first made a name for himself in 1923 by publishing a treatise on anilin and its derivatives (24) which, as a matter of fact, was reviewed rather favorably by Shreve (25). Shreve enjoyed the way in which Groggins had displayed the <u>practice</u> of chemistry within industry, and he also liked the manner in which <u>fundamental principles</u> of chemical engineering had been covered (25). However, these fundamental principles were not what could be expected - namely Little's unit operations - rather they corresponded to <u>classes of organic reactions</u> such as sulfonation, condensation, oxidation, and so on. Later, these classes of organic reactions would provide the basis for unit processes.

In retrospect, it is not too difficult to understand what Groggins had done. Being one of the very first to write a treatise on organic technology, he had structured his book according to the various <u>kinds</u> of reactions that could be relevant for the industrial handling of anilin and its derivatives. In doing so, he had transposed the organization of textbooks in organic chemistry. Because he was writing for a technical audience, he had unwittingly focused the attention of chemical engineers on <u>chemical reactions</u> and, as a result, he had made a crucial step toward the elaboration of unit processes. By underscoring this aspect of Groggins' work, Shreve's review also contributed to the evolutionary process leading to this new notion.

The term "unit process" appeared for the first time in an article on nitration penned by Groggins in 1928 (26). Retracing the precise industrial context that led Groggins to this particular article is quite interesting. Good insights are provided into the new American chemical industry and the need to reconvert productive capacity after the war. The nascent dye industry was not the only factor affecting the theoretical evolution of chemical enginnering.

Groggins' 1928 article is actually a sequel to another which was published two months earlier and which was titled "Resins from Chlorinated Cymene" (27). The key word in this title is "resins". To understand the role of resins, it is necessary to step back a little and look at wider portion of the industry.

During the war, voracious needs for smokeless powder had considerably boosted the production of pyroxylins, so that, after the war, attempts were made to divert the use of pyroxylins away from war materials toward peaceful applications. Celluloid, artificial silk, and lacquers all benefitted from the advent of peace. Lacquers, in particular, started eliciting considerable interest when it was realized that the then fast-growing automobile industry could considerably reduce the time needed to coat the body of a car with a permanent finish. Lacquers were shinier and lasted longer than traditional paints and, moreover, they could be applied within 48 hours instead of the two weeks required by the older coatings (28). But, as in the dye industry, the lacquer industry required a broad set of associated organic compounds that were, as the years went by, synthesized in the United States rather than imported from elsewhere. For example, camphor was used as a plasticizer of lacquers and until the 1920s was imported from Japan. The natural product, however, was progressively replaced by its synthetic equivalent. Other compounds with similar or even better properties than those of natural compounds were sought (28). The same was true of resins imported from tropical areas. Groggins' investigations aimed at replacing these resins with domestic synthetic products.

As a good organic chemist, Groggins paid attention to the taxonomy of chemical reactions. As a good engineer, he did not neglect the equipment needed to carry out the favored reactions and in particular the design of reactors. Both of these dimensions were very much present in his 1928 article on nitration, as they were also present in an article on a different unit process that appeared the following year: amination through ammonolysis (29). As a result, the notion of unit process gained clarification and the corresponding engineering practice was also better defined. Confronted with unit processes, the chemical engineer had to control the chemical parameters of the reaction he sought to bring about (temperature, concentration of reagents, etc.) as well as the physical characteristics of the reacting milieu (agitation, for example). The aim was to achieve the best possible yield from the chemical transformation. This of course points to the essential difference between inorganic and organic chemical industries. In the former, reactions, as a rule, either proceed completely or not at all. In the latter, slow reactions balanced by reverse reactions lead to complex questions of equilibria and incompletion of chemical processes. Inorganic chemistry does not necessarily avoid these kinds of difficulties, but inorganic industries generally incorporate the simplest of all inorganic reactions. By contrast, organic chemistry nearly always involves complex equilibria and mixtures. As soon as the industrial production of organic compounds was attempted, very difficult problems of yield, separation, and purification had to be resolved.

Going back to Groggins, his research between 1929 and 1933 displays a growing sense of what his target ought to be. In the latter year, he published what amounts to extracts of his future book on unit processes in a series of articles dealing, once again, with amination through ammonolysis (30,31,32,33,34) This series of articles is a thorough study of one process - amination through ammonolysis - from beginning to end, including the design of the reaction and economic considerations tied to the recovery rate of ammonia once the amination process is achieved. The following year, Groggins' treatise, Unit Processes in Organic Synthesis, appeared (35).

With D. B. Keyes, we reach the third and last important character on this particular stage. His part is crucial but complex because, with greater clarity than Shreve or Groggins, he perceived the intimate link between unit processes and physical chemistry. He underscored the importance of chemical equilibria for the understanding and control of industrial organic reactions (36). Keyes' role goes beyond the purely theoretical aspects of the new notion. It would not be exaggerated to state that between 1932 and 1939 he propagandized on behalf of unit processes. In 1932, he organized a symposium on the design, construction, and operation of chemical reactors (37). Then he published a series of articles on this topic. It was in the course of the 1932 symposium that he launched a campaign in favor of unit processes.

> Chemical engineering comprises not only the unit operations but also the chemical unit processes fundamental in chemical industry. Unit operations are almost entirely physical in nature - for example, distillation, filtration, grinding, crystallization, etc. Chemical unit processes on the other hand, are the common standardized processes used in the chemical industry - for example, oxidation, reduction, halogenation, hydration, nitration, esterification, etc.

> Much has been written in recent years about unit operations, but very little about unit processes....It is strange that this particular subject which means so much to every chemical engineer should have received scarcely any consideration in our modern chemical literature (38).

In his other articles, Keyes provides his readers with a picture of the U.S. chemical industry which agrees very well with the industrial context deduced from Groggins' line of investigation. The enormous surplus of nitrocellulose, ethyl and butyl alcohol, and acetic acid had to be used and the best way to dispose of them was to make them react with one another in some useful manner. The esterification of the alcohols led to solvents which could be associated with nitrocellulose to derive new kinds of lacquers - an area which started growing considerably when it was linked to the needs of a fast growing automobile industry (39, 40). The obstacle blocking this obvious conversion of a war industry to peaceful activities lay with the industrial mastery of esterification processes. As a result, this class of organic reactions became the focal point of much engineering attention after the war. Keyes drew up a long list of patents related to this particular problem, many of which dated back to the early 1920s. He also had published on lacquers in 1925 (41,42).

A few last details may be given about Keyes, which should further demonstrate the crucial importance of the evolution of organic industries immediately after the first world war. After studying under G. N. Lewis, Keys focused not only on lacquers, as we have seen, but also on synthetic dyes. Later on, he moved toward petroleum chemistry and specialized in problems of molecular cracking because he also had a good deal of experience in distillation processes while he worked for the U.S. Industrial Alcohol Company of New York in the early 1920s (43).

In conclusion, a rapid examination of the circumstances leading to the appearance of the phrase "unit processes" clearly shows that the phrase was designed in response to the brutal growth of organic industries during the war and the need to recover toward peaceful activities in the decade following the war. This overwhelming industrial demand for ways to solve innumerable problems led, among other consequences, to the invention of a new notion, although a number of chemical engineers felt that the chemical domain did not concern them.

V. The Career of Unit Processes

The relationship between unit operations and unit processes can be simply stated, at least in terms of objectives. The coining of a phrase such as unit process, clearly demonstrates the desire to create a notion both parallel and complementary to unit operations. It is on the basis of these two criteria that we can judge whether the promoters of unit processes were able to secure a safe niche for their new notion.

In an important article published in 1934, Theodore R. Olive offered a taxonomy of unit operations that sought to be as systematic as the notion would allow (44). He explicitly excluded the thorny problem of unit processes but called on someone else to solve this question (44). In response to this call (which could also be read as a challenge), Shreve drew up a taxonomy of unit processes in 1937 (45). A comparison between these two classifications quickly reveals the difference that separates the apparently analogous notions.

Olive submitted unit operations to an analysis he considered to be logical. This analysis led him to draw eight large classes covering all unit operations.

These eight categories are :

1. Handling of solids
2. Handling of fluids
3. Disintegration
4. Heat transfer
5. Mixtures
6. Separation
7. Control
8. Various physical operations which defy any rational attempt at classifying them!

Despite the fact that this taxonomy ends with a category worthy of Borges' whimsy (46), it is not difficult to see how Olive proceeded. Mentally he followed a complete productive process: substances are physically brought together (transport and size of particle); heat is applied or removed; a chemical reaction presumably proceeds; and this reaction yields a mixture which must be separated. Control is needed to keep the process going. Consequently, Olive's repeated claims of achieving a logical classification amount to little more than a systematic description of the various phases of a productive process. Even the systematic side of the taxonomy must be questioned because the eighth category is designed as a catch-all bag. Olive seems nevertheless convinced that his work, although incomplete and imperfect, can form a valid stage on the way to a completely rational classification of unit operations.

Shreve's classification of unit processes is very different and actually very brief: seven broad types of reactions suffice to cover the whole set of unit processes. He then devotes the overwhelming part of his classification to the series of variables needed to undertake the systematic study of a given unit process. In other words, Shreve seems to be treating a taxonomy as if it were a kind of checklist needed to operate a complex device. Actually, the checklist does not aim at operating anything; it is supposed to play an important pedagogic role by allowing a student to analyze any unit process in a completely systematic fashion (47) - a teaching function completely absent from Olive's classification. As a result, Shreve's response to Olive's challenge appears to miss the mark, but is offered as if it were a genuine taxonomy anologous to Olive's. This is puzzling and needs some clarification. To do so, however, we must backtrack a little.

Earlier in this text, we noted that unit processes were met by a number of attitudes ranging from indifference to chemical problems to the sentiment that chemical laws were beyond the engineer's control. But as early as 1933 D. B. Keyes had found an interesting answer to these criticisms. As if he had already heard the critique which Ittner was to voice three years later, Keyes said:

> Of course, it can be argued that these so-called unit processes may be split into the physical unit operation with modifications made necessary by chemical reactions. It should be remembered, however, that all of the physical unit operations can in turn be split into two subjects: 1) material flow, and 2) heat flow. It is more satisfactory from a pedagogical standpoint no to reduce these processes and operations to their ultimate subdivisions, principally because the lack of data handicaps the presentation of the subject in such a manner (48).

In other words, Keyes was conscious of the fact that "so-called unit processes", are not as unitary as they seem to be. Like unit operations, these processes could all be reduced to heat and material "flows". But quantitative treatments of complex reactions were unattainable at that time, and talking in terms of ultimate heat and material flows could not help teaching or solving concrete industrial problems. As a result, Keyes saw advantages in retaining notions such as unit operations and unit processes even though their claims to being fundamental units of knowledge do not hold up under close scrutiny. In effect, Keyes defended unit operations and unit processes on pragmatic grounds. The notions are ultimately to be condemned and replaced by more fundamental frameworks of analysis, but, meanwhile, they could play a number of useful roles. What remains to be examined are the factors that, in a sense, slowed down the logically unavoidable resolution of unit operations and unit processes into the more general and fundamental questions of heat and material flow.

At the beginning of this study, we recalled the tight connections linking the notion of unit operations with university curricula. We also recalled that the process of accreditation in a real sense succeeded in standardizing chemical engineering education by the late thirties. We also noted that the emergence of unit operations corresponded to older needs of the American chemical industry and that the drastically different commercial context brought about by the first world war had completely modified that industry's complexion. Unit processes were designed to respond to these new industrial needs, but the patterning of this notion as rigorously parallel as well as complementary to unit operations can also be interpreted as a move to bolster the nascent chemical engineering discipline orginally shaped by A. D. Little and his M.I.T. colleagues. Unit processes, as the term seems to indicate, merely try to extend the already powerful reach of unit operations to organic industries at a time when these industries could no longer be ignored.

However, a significant difference exists between unit operations and unit processes. The difference is the enormous complexity of the various phenomena covered by the latter of these two notions. Chemical reactions in an industrial setting are never simple and, consequently, do not lend themselves easily to quantitative treatment. This point was made repeatedly, even by promoters of unit processes, and brought with it a consequence of crucial importance. Although unit operations could both serve teaching and designing functions, unit processes could rarely address the latter. As a result, unit processes were largely confined to the classroom where they offered a systematic approach to an otherwise bewildering array of chemical reactions. Examples of their transfer to the plant were rare, even though many chemical engineers longed for quantitative models. This limitation explains why chemical engineers treated unit processes as a kind of heuristic device at best, and why unit processes remained confined to teaching situations in most cases. It also partially accounts for the pedogogic bias already noted in Shreve's attempt to create a taxonomy of unit processes.

Embedded in a firmly entrenched, standardized curriculum, both unit operations and, to a lesser extent, unit processes, were provided with a kind of historical resilience far beyond the reach of their simple conceptual structures. In the case of unit processes, their pedagogic function partly legitimized their continued existence for at least two or three decades. A situation made relatively stable because the theoretical front of chemical engineering did not witness any significant advances. In other words, as long as theoreticians were unable to quantify models of complex organic reactions, unit processes could retain a certain raison d'être. In effect, the conceptual imperfection unit processes had to be accepted as long as there was nothing better to replace them. This does not do full justice to unit processes. Actually, their introduction within chemical engineering coincided with the increase of theoretical and quantitatively oriented research on chemical reactions. Equilibria, industrial stoichiometry, and kinetics were the objects of much solicitude on the part of chemical engineers. Unit processes were often viewed by chemical engineers, including D. B. Keyes, as the poor cousin of unit operations. Lack of interest was not the cause of this attitude, it was the enormous difficulties associated with the domain. The presence of unit processes in chemical engineering was a temporary solution to problems that could already be stated with clarity in the 1930s, but could not be solved.

Other factors also provided a degree of added stability to unit processes. In particular, the analytical "taxonomy" offered by Shreve in 1937 not only played an important role in the classroom, it also provided a useful guide for designing a chemical reactor whenever a series of approximations had to replace the equations ruling a well-defined model - and that was most of the time (49).

VI. The End of Unit Processes

In 1955, the fourteenth volume of the Encyclopedia of Chemical Technology appeared containing a fairly substantial article on both unit operations and unit processes. Predictably, the author of this entry was R. N. Shreve, and the story he told was the now familar chronology mentioned earlier. Subtle shifts in vocabulary, however, signal the growing inability of unit processes to maintain their unitary characteristics. On the one hand, Shreve noted that the fundamentals of unit processes dealt with kinetics, yields, and problems of conversion, all of which refered to the increasing importance of concepts borrowed from physical chemistry, as predicted by D. B. Keyes more than two decades earlier. On the other hand, Shreve was forced to acknowledge that the term "unit process" covered a wider variety of events than its physical counterpart "unit operations". His new table of unit processes included twenty-seven categories of processes starting with combustion and ending with ionic exchange. But Shreve argued, "... the unitary concept, as applied to physical and chemical changes, has been useful and has emphasized the fundamental systems and principles rather then the technical details" (50). In so arguing, Shreve implicitly reiterated the pedagogic value of unit processes while admitting that the concept was of little value to the engineer in the plant confronted with technical details that he must keep under control.

In a sense, Shreve's 1955 article was the swan song of unit processes. As early as the following year, he wrote that the expression "unit process" was being replaced by the more neutral term "chemical conversion" (51). That is to say, the unitary dimension of unit processes so dear to Shreve, Groggins, and

Keyes had to be abandoned. The reasons for this evolution will not be examined here because they would require another paper to be treated satisfactorily. A look at the second edition of the Encyclopedia of Chemical Technology provides enough information to formulate one basic hypothesis. In the second edition, no entry appears for unit operations and unit processes. It is as if these two notions had never existed. A new entry, however, had been added - "Transport Processes" - and it takes little time to see that D. B. Keyes' prediction had at last come true. Transport processes, or phenomena as they were defined in this encyclopedia, dealt with all the artifical or natural systems where physical or chemical transformations took place. As far as engineering is concerned, this extremely general notion refers to the analysis and design of systems where moment, energy, and mass transfers occur (52). In other words, transport phenomena deal with all the questions addressed by unit operations and unit processes. In so doing they marshall the most basic concepts granted by physics and physical chemistry. Intermediate, apparently unitary, notions have but little to add to this new and extremely general formulation of chemical engineering problems.

VII. Conclusion

In the history of science, as in the history of technology, incomplete successes or outright failures have attracted little attention. This neglect may be because historians see few advantages in linking their names to a forgotten episode. Also, such episodes are often more difficult to handle than success stories precisely because the main interest of such stories almost always lies behind or beside the ostensible object of study.

The unit process story readily demonstrates this latter point. First appearing as a the partially unsuccessful attempt to extend the operational capability of unit operations to the chemical domain, it evolved as a response to new industrial demands, particularly in the organic domain. The grafting of the new notion on the main trunk of chemical engineering as defined by unit operations was operated within the professional and educational confines of the domain. But, because this grafting never really took, what is the point of retrieving such an ambiguous string of events?

The answer to this question lies in the fact that unit processes shed a very interesting light on unit operations. First of all, the very existence of unit processes demonstrates that, paradoxically, chemical engineering, as structured by unit operations, could not easily address the purely chemical aspects of its domain.

A second point can be made about the way in which unit processes came into existence. The emergence of unit process is analogous to that of unit operations, particularly with regard to relevant social groupings involved in this evolution. Industries, professional organizations, particularly AIChE and the American Chemical Society, and universities are involved, as in the case of unit operations. This time, however, the process leading to the identification of unit processes is simpler because the disciplinary identity of the chemical engineering profession is well defined and especially because unit processes, as the name shows, are patterned after unit operations.

At this point the really interesting results start appearing. First of all, as noted earlier, unit processes, despite appearances, are not as parallel and complementary to unit operations as they are generally presented. The discrepancy between Olive's classification of unit operations and Shreve's classification of unit processes amply demonstrates that, with the best intentions in the world, the supporters of unit processes did not succeed in treating unit processes exactly like unit operations. This failure occurred because no one succeeded in developing adequate quantitative models for the treatment of chemical reactions as had been done in the case of the physical handling of materials. No one could approach a hydrogenation reaction in the way one could treat a distillation tower, for example. As a result, unit processes stood more as an unfulfilled desideratum than as a solution because they offered almost no help to the practicing engineer. Only in the classroom did unit processes offer a way to organize a vast collection of facts which otherwise would have had to be presented without even a pretense of order. Their superficial analogy with unit operations allowed unit processes to play this organizational role in the classroom, but not in the plant.

The failure to lend itself to good quantitative modelling goes a long way toward explaining why unit processes ultimately failed to establish themselves permanently in the discipline. This failure was eventually of some importance because its causes were progressively recognized and identified. These causes involved insufficient development of physical chemistry, insufficient means of computation and the like. Moreover, the general form of the solution to be sought was also identified rather early. It would involve recasting the whole question of chemical reaction engineering in terms of heat and material flows. But when D. B. Keyes pointed this out, he also pointed out - and that is crucial - that unit operations would also be dissolved, so to speak, in this far more general conceptual scheme.

In Keyes' mind, the fundamental solution he outlined was unattainable for many years and even decades, and the evolution of chemical engineering demonstrated the validity of his reasoning. Consequently, he thought it was much better to adopt a pragmatic attitude and hold on to unit operations and unit processes for as long as they retained some degree of use to teach, to design, or to manage industrial processes.

As a result, the limited ability of unit processes to create a viable niche for themselves within chemical engineering must ultimately be understood in terms which also involve unit operations. Although the historical resilience of unit processes turned out to be less than that of unit operations, it was no different in its essential elements. Studying the uneasy and ultimately unsuccessful career of unit processes can therefore be easily justified as a way to shed light on the far more successful career of unit operations. In particular, the career of unit processes raises a hypothesis about the evolution of unit operations. The staying power of unit operations was not so much because of the structural coherence of its conceptual elements as its essential links with social and, more specifically, professional groups. As a theoretical entity, unit operations appears far less stable and, in fact, appears quickly threatened by notions which rest on fewer and more fundamental scientific concepts. Ultimately, this threat came to be realized with the advent of transport phenomena, but this is another story. In effect, unit processes can be interpreted as both the attempt to extend the reach of unit operations and a symptom of their conceptual fragility.

Literature Cited

1. Guédon, J.-C. Testie Contesté 1981, 5, 5-27.
2. Servos, John Isis 1980, 71, 531-49.
3. Little, A. D. Chem. Metall. Eng. 1921, 24, 1047-53.
4. Parmalee, H. C.; Trans. Am. Inst. Chem. Eng. 1932, 28, 322-3.
5. Guédon, J.-C Testie Contesté 1981, 5, 5-27.
6. Bucholz, K. Social Studies of Science 1979, 9, 33-62.
7. Shreve, R. N. "Chemical Process Industries"; McGraw-Hill: New York, 1945.
8. Kirk, R. E; Othmer, D. F. Encyclopedia of Chemical Technology"; Interscience Encyclopedia, Inc.: New York, 1955; Vol. 14, pp. 422-5.
9. Shreve, R. N. Ind. Eng. Chem. 1948, 40, 379-81.
10. Trans. Am. Inst. Chem. Eng., 1935, 32, 568.
11. Shreve, R. N. Ind. Eng. Chem. 1940, 32, 145.
12. Chilton, T. H. Ind. Eng. Chem. 1951, 43, 295.
13. Chem. Metall. Eng. 1923, 29.
14. Keyes, D. B. Trans. Am. Inst. Chem. Eng. 1936, 32, 491-2.
15. Whithrow, J. R. Trans. Am. Inst. Chem. Eng. 1936, 32, 490-
16. Flett, L. M. Industrial Eng. Chem. 1951, 43, 305.
17. Shreve, R. N. J. Ind. Eng. Chem. 1918, 10, 789-90.
18. Browne, C. A.; Weekes, M. A. "A History of the American Chemical Society"; American Chemical Society: Washington, D. C., 1952; 278.
19. Schroellkopf, J. J. Ind. Eng. Chem. 1918, 10, 792-3.

20. Matos, L. J. J. Ind. Eng. Chem. 1918, 10, 791.
21. Schroellkopf, J. J. Ind. Eng. Chem. 1918, 10, 793.
22. Flett, L. M. Ind. Eng. Chem. 1951, 43, 304.
23. Olive, T. R. Chem. Metall. Eng. 1934, 41, 229-31.
24. Groggins, P. H. "Anilin and its Derivatives"; Van Nostrand: New York, 1924.
25. Shreve, R. N. Ind. Eng. Chem. 1924, 16, 265.
26. Groggins, P. H. Chem. Metall. Eng. 1928, 35, 466-7.
27. Groggins, P. H. Ind. Eng. Chem. 1928, 20, 597-9.
28. Calvert, R. J. Chem. Educ. 1925, 2, 359-74.
29. Groggins, P. H. Chem. Metall. Eng. 1929, 36, 273-5.
30. Groggins, P. H. Ind. Eng. Chem. 1933, 25, 42-6.
31. Ibid., 46-9.
32. Ibid., 169-75.
33. Ibid., 274-6.
34. Ibid., 277-9.
35. Groggins, P. H. "Unit Processes in Organic Synthesis"; McGraw-Hill: New York, 1935, 1938, 1947, 1952, 1958.
36. Keyes, D. B; Lewis, G. N. J. Am. Chem. Soc. 1918, 40, 472-8.
37. Ind. Eng. Chem. 1932, 24, 1091-1109.
38. Keyes, D. B. Ind. Eng. Chem. 1932, 24, 1091.
39. Ibid., 1096.
40. Little, A. D. J. Chem. Educ. 1928, 5, 646-7.
41. Keyes, D. B. Ind. Eng. Chem. 1925, 17, 558-67.
42. Keyes, D. B. Ind. Eng. Chem. 1925, 17, 1120-2.
43. Keyes, D. B. J. Chem. Educ. 1938, 15, 480.
44. Olive, T. R. Chem. Metall. Eng. 1934, 41, 229-31.
45. Shreve, R. N. Ind. Eng. Chem. 1937, 29, 1329-33.
46. Foucault, M. "Les Mots et les Choses"; Gallimard: Paris, 1966; 7.
47. Shreve, R. N. Ind. Eng. Chem. 1937, 29, 1330.
48. Keyes, D. B. Chem. Metall. Eng. 1934, 41, 244.
49. For example, Shreve, R. N. Ind. Eng. Chem. 1940, 32, 146.
50. Kirk, R. E.; Othmer, D. F. "Encyclopedia of Chemical Technology"; Interscience Encyclopedia Inc.: New York, 1955; pp. 422-5.
51. Shreve, R. N. "Chemical Process Industry"; McGraw-Hill: New York, 1956; 3rd Ed., 1-21.
52. Kirk, R. E.; Othmer, D. F. "Encyclopedia of Chemical Technology"; John Wiley & Sons: New York, 1969; 2nd Ed., Vol. 20, p. 610.

RECEIVED May 19, 1983

4

Will Milk Make Them Grow?
An Episode in the Discovery of the Vitamins

STANLEY L. BECKER
Bethany College, General Sciences, Bethany, WV 26032

Following the enunciation of the vitamin hypothesis in 1912, the value of cow's milk as a source of vitamins necessary for normal growth of rats (and, by implication, for normal growth of humans) became a significant issue in animal feeding experiments. While there was general agreement about the qualitative value of milk as a vitamin source, a major conflict arose between two pioneering research laboratories concerning the quantitative aspects. Although the researchers involved reported differences in quantitative requirements that varied by as much as 5-8 fold, the disparity between results was never resolved and remains a mystery to this day.

Critical Events

June, 1912: Casimir Funk, a Polish biochemist working in London, England, wrote a review article in which he elaborated on the fact that diseases such as scurvy, beri-beri, rickets and pellagra had long been known to be associated with the dietary. Funk said that these diseases could be prevented or cured by adding certain organic substances to the diet, substances he called vitamines. (1)

Mid-July, 1912: Frederick Gowland Hopkins, a biochemist at Cambridge University, published an elaborate paper demonstrating that additions of small amounts of fresh, whole milk to otherwise deficient diets of experimental rats were followed by periods of normal growth and development of the animals. (2)

July, 1913: E. V. McCollum, a biochemist at the University of Wisconsin Agricultural Experiment Station published a paper in which he showed that there was something in butter that made rats grow; someting that was later to be named vitamin A and, later yet, vitamins A and D. (3)

August, 1913: T. B. Osborne, a protein chemist at the Con-

0097-6156/83/0228-0061 $06.75/0

necticut Agricultural Experiment Station, in collaboration with L. B. Mendel, a physiological chemist at Yale University, published a paper in which they demonstrated that butter was very effective in making rats grow. (4) This work had been preceded by a monograph in 1911 which contained a wealth of data on the nutritional value of isolated food substances as related to the growth of rats. (5)

In December 1929, Frederick Gowland Hopkins shared the Nobel Prize in Medicine or Physiology, "for his discovery of the growth-stimulating vitamins."

The story related here will clarify, to some extent, the roles of these pioneers in the emergence of the vitamin hypothesis. More importantly, it will shed light on some peculiarities of nutritional biochemistry as it was practiced in the first quarter of the twentieth century. In particular, it will focus upon experimental results that helped one investigator win a Nobel Prize, results that could not be reproduced by some of the most competent research workers in animal feeding experiments.

Nutrition, 1905 Style

The dominant principle of nutrition in 1905 was energy. Calories were counted for people in different occupations, consuming highly varied diets. Given sufficient calories, nutritional dogma said, and human beings will grow and develop normally. The calories could best be obtained from fats and carbohydrates. In addition, a minimum amount of protein was required (and this minimum was hotly debated) along with a few minerals and water.

Much has been written about these early twentieth century views of nutrition, both of the rigorous scientific variety and of the less formal but widespread folk concepts. The pre-history of the episode being related here is reasonably documented in a number of works such as those of McCollum (6) and Becker (7). The literature shows clearly that in the first decade of this century, calorie-dominated nutrition was being quietly but vigorously challenged. In the second decade the challenge became open and even more vigorous.

Pioneers

Casimir Funk had come from Poland to the Lister Institute in London, England to investigate the chemical nature of the substance in rice-polishings (rice bran) that cured or prevented polyneuritis in birds, a disease closely related to beri-beri in humans. His research in this subject coupled with his awareness of what were becoming known as diet-related diseases, the deficiency diseases, led to his review article of 1912 in which he coined the term, vitamine. Vitamine, to Funk, was actually a variety of chemical substances, organic amines, that acted selectively in the prevention or cure of diseases such as beri-beri, scurvy, pellagra and other diet-associated pathologies. In short, they were vital amines, hence vitamines.

In the mid 1920's, Funk became embroiled in a controversy over the discovery (and the discoverer) of the vitamines, or vitamins as they were then known. (8, 9, 10) He gave some credit to Hopkins for his early endeavors but refused to acknowledge Hopkins' priority. This episode bears strongly upon remarks made by Hopkins in his Nobel Address of 1929 wherein he discussed Funk's work as well as McCollum's and that of Osborne and Mendel.

Frederick Gowland Hopkins, analytical chemist, physician and biochemist, had conducted numerous experiments in animal feeding prior to his famous comment in 1906 that no animal could live on a diet of pure protein, fat, carbohydrate, minerals and water. He cited the simple fact that animals live upon plants or other animals whose tissues contain many other substances besides those usually considered adequate for a normal diet. "...it is certain that there are many minor factors in all diets of which the body takes account." (11)

These comments were made near the end of a talk at a meeting of the Society of Public Analysts, a group composed principally of analytical chemists and medical doctors. During the discussion which followed, almost no reference was made to Hopkins' new ideas and remarks, in itself a measure of the newness of such ideas along with the inability or ignorance or lack of attention paid to such comments by the group in attendance.

Over the next six years Hopkins pursued, intermittently, the search for the other nutrients required to keep an animal alive and in good health. In 1912 came the published results of his research and while he confirmed them in publications of 1913 (12) and 1920 (13), it appears that other pioneers in nutritional studies, Osborne and Mendel in particular, were unable to repeat or reproduce his results.

Thomas B. Osborne and Lafayette B. Mendel worked together for nearly twenty years to unravel the chemical and physiological properties of proteins. In their efforts to supply ever more sophisticated diets, i.e., diets made up of pure chemical proteins, minerals, etc., they discovered that different proteins were markedly different in their ability to promote growth or even to maintain animals at contant body weight.

Repeated failures of such diets, contrasted with successful ones on which animals grew normally when powdered whole milk was the major ingredient, forced Osborne and Mendel to examine milk itself to explain its peculiar ability to sustain growth and maintenance of experimental rats over extended periods of time. Ultimately they discovered the growth-promoting property of butter and butterfat. (14) Not surprisingly, with so many ideas "in the air" and with other researchers prowling in related areas, Osborne and Mendel were pre-empted by three weeks in publishing their discovery. It was E. V. McCollum at Wisconsin who first reported that 'butter makes them grow.'

Elmer V. McCollum, while obtaining his doctorate in chemistry at Yale, took some courses in physiological chemistry with Mendel as his instructor. McCollum also worked in Osborne's laboratory at the Connecticut Agricultural Experiment Station for a few months and was hired by the University of Wisconsin Agricultural Experiment Station as a chemist to analyze the intake and outgo of cattle on restricted rations, i.e., cattle being given food obtained from a single crop plant.

Not content with such a laborious and time consuming enterprise, McCollum started, on his own initiative in 1908, the first rat colony in the United States to be devoted exclusively to the study of nutrition. More than five years later he was able to report that a successful diet for a rat required the inclusion of an ether extract of egg or butter. (15)

Confusion, Uncertainty and Disagreement

Although the white or albino rat was to become the animal of choice in nutritional investigations, chickens, pigeons, cows, pigs, guinea pigs, rabbits, mice and dogs (and probably a few cats) were also used in an attempt to better understand the physiological properties of foods. Species' differences in dietary needs were only slowly recognized, a situation that led to much confusion, uncertainty and disagreement over the results and significance of experimental work. What kind of sense, e.g., could be made of the fact that guinea pigs and human beings could develop scurvy while rats could not? Why was it so difficult to induce deficiency diseases in cattle? Was pellagra confined to humans alone?

These were some of the questions that faced nutrition researchers in the period of primary concern in this paper, 1910-1925. Other problems added to the high level of doubt and skepticism that prevailed. Casimir Funk, e.g., in isolating what he thought to be the amine necessary for the control of beri-beri, also obtained nicotinic acid from rice polishings and from yeast, both excellent curatives/preventives of beri-beri. (16) Even though Funk had suggested that each deficiency disease was the result of the lack of a specific chemical entity, he never realized in 1911-1935 that the pellagra cure/preventive, nicotinic acid, was sitting in a chemically pure state in his test tubes.

Hopkins was late in realizing that one of his 'accessory food substances', later called vitamins, was contained in the fat of milk as well as in the fats of other animals and in certain plant tissues as well. (17, 18) Osborne and Mendel actually reached a point in their research in 1912 that made them confident that fats as fats were not necessary at all in the diets of experimental rats. (19) McCollum went so far as to state at one time that aside from the anti-scorbutic substance (later to be known as vitamin C), there were only two new substances to be concerned with in rat nutrition: 'fat-soluble A' and water-soluble B'. (20) McCollum made it very clear in this 1916 paper that he did not like the word vitamine (which, nevertheless, became vitamin in 1920). (21)

The conflicting results and various interpretations obtained from animal feeding experiments were, of course, related to many factors, a point that can be illustrated in various ways. The author has selected what he calls the 'milk-is-good-for-you-but-how-much-is-enough?' problem that emerged, for the most part, in the investigations of Hopkins at Cambridge and Osborne and Mendel at the Connecticut Agricultural Experiment Station in New Haven during the interval, 1911-1922.

Milk is Good for You But...

In July 1911, Osborne and Mendel completed the second part of the monograph cited earlier, (22) published by the Carnegie Institution which had partially supported Osborne's researches for a number of years. This monograph of 138 pages, containing many tables and 129 charts, was the most comprehensive document published to date on animal feeding experiments. It represented the data, rationales and conclusions of nearly two years and illustrated vividly Osborne and Mendel's major premise: different proteins, either individually or in combination, were remarkably different in their abilities to either maintain a rat at constant weight or to allow marginal or even normal growth to occur in these animals.

Osborne and Mendel were able to make this premise in part because they had develped two diets that allowed rats to grow normally: the mixed food (dog biscuit, sunflower and other seeds, fresh vegetables and salt) and the milk food (whole milk powder, starch, a mixture of mineral salts and lard). It was the latter dietary that, when comprising all or only a part of an experimental ration, would prove to be a source of much confusion about the nutritional value of milk.

In July 1912, Hopkins had published his first paper concerned with feeding small amounts of whole milk to rats on experimental diets. In August he wrote separate letters to Osborne and Mendel. The following is from the letter to Osborne:

> I trust that neither you nor Prof. Mendel will think that my paper which has just appeared in the Journal of Physiology (of which I am now sending you a reprint) intrudes unduly into your domain. As a matter of fact I have engaged in such feeding expts. for a good many years, and the great pleasure with which I read your work...was (human nature being what it is) somewhat spoilt by the fact that it takes the wind out of the sails of many scores of experiments of my own! (23)

Hopkins stated further that he was able to obtain growth in rats on diets composed of amino acid mixtures only when small amounts of "yeast-extract etc. were added." He said nothing whatsoever about the milk supplements which underlie his entire 1912 paper.

This being summer, Osborne and Mendel often went to their vacation retreats from where they carried on an extensive correspondence. In one of these letters, Mendel remarked to Osborne:

> F. G. Hopkins wrote me a pleasant note, as he presumably did to you. I have not yet seen the reprint which he states he has sent, but I shall reply that we do not regard ourselves as the sole scientists entitled to feed rats. Evidently he is impressed with the idea of a "growth substance" or something of that sort, aside from the ordinary nutrients. (24)

Nearly two weeks later, after reading Hopkins' paper, Mendel wrote again to Osborne:

> The Hopkins paper was very interesting....Of course his foods are all comparatively crude; and our positive exp't with fat-free art. p. f. m. is hard to understand except on the assumption of a store of the growth factor of the body. ... The energy proposition gets a great set back by Hopkins work and our growth paper. ...However, these 'accessory' extracts may be an invaluable aid to us in studying the comparative role of proteins if they permit us to keep our animals in better form. The point in our paper which Hopkins questions is very important i.e., the quantity of p. f. m. that is sufficient, and it needs to be repeated perhaps. (25)

Explanation of two terms mentioned above is in order. Osborne and Mendel had been successful in growing rats on the milk food diet. In order to have all the advantages of milk without the protein content, they developed what they called protein-free milk or p. f. m. Since their primary research concern lay with proteins and because milk was a complete food by itself, they assumed that if the water and proteins (mostly casein) were removed from milk the resultant product would have all the characteristics of milk while allowing the experimenters to introduce other proteins into animal diets. Osborne and Mendel would thereby be able to evaluate the nutritive value of any protein when it was fed with p. f. m. (26)

Art. p. f. m. or artificial protein-free milk, was a logical next step in their work. Since they were attempting to grow rats on totally synthetic diets, i.e., diets composed of pure chemicals, Osborne and Mendel attempted to duplicate the composition of p.f.m. using laboratory or commerical grade chemical salts and lactose in combinations that approximated as closely as possible the composition of natural p. f. m. (27)

The Connecticut researchers had great success using these protein-free adjuvants but only up to a point. While extensive growth might occur especially with the natural p. f. m., sooner or later the rats stopped growing. Some could be maintained at nearly constant weight for months but inevitably all rats would begin to decline on diets that contained p. f. m. or art. p. f. m. As Osborne wrote to Mendel: "It is becoming more and more evident that they [the rats] get some essential substance from the milk, the nature of which we do not yet know." (28)

...How Much is Enough?

Although Osborne and Mendel were coming to terms with the probability that 'some essential substance' was present in milk, they had already that year (1912) published two articles which led readers to believe that Osborne and Mendel had developed a totally synthetic and successful diet. In early 1913, Hopkins and Neville questioned the validity of some of the experimental results obtained by Osborne and Mendel. They said that the Americans' results seemed to indicate that young rats could grow on purely artificial diets and, therefore, that accessory factors or vitamines were dispensable. "But...they contradict what is now a considerable body of experience, and the experiments which yielded them seem to call for repetition." (29)

Hopkins and Neville went on to say that they had fed rats on the diet used by Osborne and Mendel with one significant difference: they extracted the protein and starch with alcohol and recrystallized the lactose (milk sugar) from an aqueous solution by addition of alcohol. The methods of feeding were the same as those described by Hopkins in his 1912 paper. The results? Rats ceased to grow, all before 15 days on the extracted diet. After weight decline set in some of the rats were given 2 cc of milk per day per rat following which growth resumed and health was maintained.

The authors also stated that their rats behaved differently from the three animals fed by Osborne and Mendel but that the difference could not be explained. Hopkins had found in earlier experiments that very small amounts of substances extracted with hot alcohol from foods would, when added to deficient diets, render them capable of some growth potential. It was apparent, to Hopkins at least, that hot alcoholic extractions of foods removed small amounts of effective growth promoter from the diet. Ether extraction as carried out by Osborne and Mendel in their fat-free experiments was considered by Hopkins to be inferior to alcoholic extraction. "The purpose of the present note is to indicate that there is still reason for a continuance of the search for special accessory substances of potent influence upon growth. It should be pointed out that Osborne and Mendel themselves admit that such substances may exist." (30)

In late February of 1913, Osborne and Mendel responded to the note in the Biochemical Journal in a letter to Hopkins in which they stated that they had been surprised by the outcomes of the experiments on purely arificial diets and that newer experiments demonstrated either poor growth or none at all on fat-free artificial diets. They also said in this letter that they believed they had a clue as to the cause of the differences in results. Furthermore:

> All of us are dealing with problems which involve extremely complicated conditions, and we must be prepared to be confronted with unexpected results for some time to come. For example, our experience in studying the effect of small additions of milk to various artificial diets has rarely been that which we expected from your report....We have no intention of intruding on your field in making these experiments, nor do we expect to publish our results as evidence that your experiments involve error, or that your published conclusions are incorrect. ...We are convinced that the pursuit of the growth substance involves some of the most important problems of biochemistry, and that it will not be long before the combined efforts of those who engage in it will solve at least some of them. (31)

No mention was made of the alcohol v. ether extraction of foods but in their paper reporting the results of feeding fat-free diets Osborne and Mendel stated clearly that none of the foods had been extracted with hot alcohol. Furthermore, they found it difficult to believe (this was in mid-1912) that skimmed milk could contain an important lipoid (fat-like substances such as lecithin, e.g.) in any adequate amount, "while...butter, which must contain some compounds of this type [is] inadequate." (32)

Here we can see a major difference in the thinking of Hopkins and that of Osborne and Mendel. For Hopkins, very small quantities of unknown organic substances, soluble in hot alcohol, were necessary for growth. Osborne and Mendel were still clinging to the belief that effective nutrients would have to be present in quantities greater than the 'traces' implied in Hopkins' experiments. Still, Osborne and Mendel recognized that only additional research would resolve the conflicting results.

Hopkins replied to their letter on March 14, 1913. (33) He apologized for the tone of the Biochemical Journal note: "The fact is I have done so much work in this (not very successful) endeavour to separate the unknown substances which affect growth, that when your paper ...came out I suffered from an attack of nerves." (34) Hopkins went on to discuss his success with feeding small quantities of milk to growing rats, and said that the story was clear under his experimental conditions. Furthermore, he commented that the organic constituents which he labeled, 'exogenous growth hormones,' were not the same as Funk's vitamines. Hopkins was most emphatic in pointing out the need to thoroughly extract food so as to remove the growth factors. Only when the basal dietary was well extracted, he concluded, could the effects of small addenda (such as milk) make themselves visible. "That such extraordinary small amounts of the 'x-factors' can act, rather diminishes the practical importance of the phenomenon; but not, I think, its theoretical interest." (35) Eight years later Hopkins would express a different attitude toward the practical importance of the phenomenon; by then vitamins were 'very big' in dietetics.

Osborne and Mendel prepared a draft letter reply to Hopkins and it can be assumed that a final version was sent, most probably in April 1913. (36) This letter re-emphasized the problems of apparently conflicting results and the methods used to solve such problems. They said that they had come to agree with Hopkins' conclusion that some organic substance, in small quantity, was necessary for growth. The p. f. m. diets seemed to be adequate for about 80 days; the art. p. f. m. diets were highly variable in their results; the standard milk food was always adequate. "...we now think that milk contains something essential for growth which is either partly or wholly destroyed or removed by our process of making the protein free milk." (37) Osborne and Mendel also stated their belief that the young rat is born with a surplus of this substance, upon which it can grow for a period of time. Furthermore, they confirmed that addition of small quantities of alcoholic extracts of some foods did indeed contain a growth promoting substance although they did not know whether it was organic, inorganic or a combination. And yet, they said further on that alcoholic extraction of the proteins could not be significant in their work since their results were controlled by using the same proteins in diets which were incapable of inducing growth in rats.

Interim

Less than four months after the writing of the letter to Hopkins, McCollum as well as Osborne and Mendel had observed and published the fact that butter contained something that made rats grow. In this same year, 1913, Hopkins became involved with the administration of the Medical Research Committee, (M. R. C.), an arm of the new National Insurance Program in Great Britain. As a consequence, Hopkins was not able to carry out any significant research on his 'growth promoting accessory factors' in milk. Nevertheless, his association with the M. R. C. ultimately brought him into international prominence and provided him (and his associates) with a forum from which he could speak to a larger audience about the growth factors, better known as vitamins.

With the outbreak of World War I in 1914, the need for practical dietetics consumed the time and labors of many a British nutritionist so that while basic research in the vitamins was conducted to a limited extent in Great Britain, it was in the United States that such research almost literally "took off."

McCollum in Wisconsin and Osborne and Mendel in Connecticut pursued their breakthroughs with many significant contributions. McCollum, in 1915, showed that there were two growth substances needed by rats, one soluble in fats, the other soluble in water or alcohol. (38, 39) In these same papers McCollum also showed that non-extracted lactose (or unrecrystallized lactose) was capable of supporting some growth in rats. Furthermore, he pointed out that commercial lactose or even some laboratory grade lactose contained small quantities of a growth-promoting substance or substances, an observation confirmed by Drummond in 1916. (40)

Osborne and Mendel continued their extensive studies of the nutritive values of proteins while extending the list of foods known to contain vitamines. (41, 42) They attempted to isolate the growth promoter in butter with no success but they made significant progress in their quest for a totally synthetic diet by developing an excellent salt mixture for the basal dietaries of their rats. (43)

Thus, at the war's end in 1918, vitamines or vitamins or growth-promoting accessories, both fat-soluble A and water-soluble B were part of the lexicon of nutritional science. Even though it would require more than another decade before the first vitamin was chemically indentified, the need for vitamins in rat growth was no longer in doubt. What is both remarkable and fascinating here is that the rats' needs were immediately translated into human needs. The war had evoked high levels of interest in human nutrition. How much of which foods were required for good health? The discoveries made in the nutritional requirements of rats only served to heighten this interest. In short, if it was good for rats it was good for humans.

Yes, but How Much Milk is Really Enough?

While both the fat-soluble and water-soluble vitamins had been found in a variety of foods, the problem of growth on milk diets had not yet been resolved, at least not in the minds of Osborne and Mendel. In a 1918 publication they said that they were impressed by the "apparent discrepancies in the quantitative relations of the amounts of milk required to furnish the vitamine factor in our experiments in contrast with those of Hopkins." (44) They went on to claim (and justly so) that a better understanding of the quantitative relations was highly desirable. "In view of the results of Hopkins' experiments it has become generally believed that milk is one of the richest sources of the water-soluble vitamines among our food products." (45) The experimental results shown in this publication were certainly not in accord with those of Hopkins in his 1912 paper.

Osborne and Mendel tried, and apparently failed, in two additional experiments to reconcile their values for the quantities of milk needed with the quantities claimed by Hopkins. (46, 47) What were these irreconcilable values? Hopkins claimed that 2 cc of fresh whole milk would suffice for a young rat of 50-100 g weight when given daily. Osborne and Mendel's determinations indicated that 10-16 cc of fresh whole milk would be needed on a daily basis. If one makes a crude estimate of the equivalent quantities necessary for young humans, using body weight as the major criterion, then a baby of 5 kg would require about 200 cc per day according to Hopkins and 1000 to 1600 cc per day according to Osborne and Mendel, just to supply the necessary water-soluble vitamin(s). (Of course, such calculations assume that no other foods containing the vitamin(s) were being supplied.)

During the winter of 1918/19, while Mendel was a member of the Allied Food Commission in Europe, he visited Hopkins in Cambridge. Upon his return the two research groups corresponded on the subject of the milk minimum necessary for a growing rat. In December 1919, Osborne and Mendel sent Hopkins a copy of a paper they were going to publish, "...presenting our newest experience with milk produced in the summer months." (48) Here, Osborne and Mendel were dealing with the possibility that summer milk was richer in vitamine than winter milk, a plausible hypothesis since dairy cattle ate very different rations in the two seasons. Hopkins was certainly alert to this possibility and in addition, it can be said that extensive research by a number of people involved with agriculture and dairying indicated that while fat-soluble A was significantly variable with respect to season, the water-soluble B appeared to be fairly constant in its concentration in milk throughout the year.

In this same letter to Hopkins, Osborne and Mendel suggested that Hopkins comment on their new paper or perhaps even publish jointly.

> We shall also be glad to have suggestions as to the best way to present our combined results to the public. Our chief concern is to avoid adding to the confusion already existing in respect to this important subject. Perhaps with the help of our newest experiments you can now put the entire matter of the growth-promoting effect of milk in its proper relations. (49)

Hopkins replied in February 1920, politely declining any joint publication and urged Osborne and Mendel to publish their work independently of his results.

> Let me say at once that my experience during the past year has convinced me that you are perfectly right in your conclusion that the small quantities of milk used in the experiments published in my paper of 1912 are not, under average conditions, capable of maintaining growth in the animal. Nevertheless I have, with cer-

> tain precaution, succeeded, in three experiments in reproducing the results of the paper in question....Clearly however, considerably more milk is required for continued growth at a heavier weight. ...You will understand that in my Journal of Physiology paper I had no idea of deciding the absolute amount of milk required to supply the demands for vitamines, but rather of emphasising, generally, the importance of accessory factors in diet. (50)

Hopkins also recounted the experiences that led him into the accessory factors problem in general, again placing emphasis upon extraction of foods with hot alcohol (80%) in order to demonstrate the loss of nutritive power of such foods when so extracted. He also said that he wished he had had the insight in those days to have recognized the existence of the two growth factors demonstrated by Osborne and Mendel and by McCollum.

(This writer would like to know what 'certain precautions' Hopkins took in order to reproduce his 1912 results but we shall return to those results shortly to see what they meant in the context of nutritional science of the day.)

In late February 1920, Osborne and Mendel replied to Hopkins' epistle in a manner that belies their activities:

> It seems to us that you attach too much importance to the quantitative relations of milk as a source of the water-soluble vitamine. Even though larger quantities of milk may ultimately be demonstrated to be necessary than your original experiments indicated, we are sure that no one will ever think of criticizing your early work on that score. The essential importance of your discovery is in no way affected. ...Pioneers in such fields of investigation are not expected to settle quantitative limits at the outset of such discoveries. ...Our work was simply designed to elaborate your pioneer investigations and define the conditions more exactly than was possible eight or nine years ago. (51)

Note that Osborne and Mendel believed that larger quantities of milk would probably be needed to furnish sufficient vitamine. They may have said that the quantitative relations were not too important but they exerted significant efforts to determine those relations. The paper they had sent to Hopkins was published in 1920 and their last attempt to discover the quantitative relations was published in 1922.

In the spring of 1921, Hopkins went to the United States where he visited, among other places, the laboratories of Osborne and Mendel as well as McCollum's now located in the School of Public Health at Johns Hopkins University in Baltimore. Shortly after Hopkins returned to England he received a note from Mendel informing him that a Japanese research group had obtained significant growth of rats given 2-3 cc of milk on an otherwise vitamin B deficient diet. (52)

In 1921 came confirmation of a sort in validating Osborne and Mendel's claim that 10-16 cc of milk were needed per day by the rat. Dutcher reported that 10 cc of milk obtained from cows on vitamine-rich rations usually sufficed in furnishing sufficient A and B for normal growth, while larger quantities of milk produced on vitamine-poor rations were very deficient in both food accessories. However, water-soluble B did not fluctuate to any great extent since most dairy animals received some grain. (53)

On the other side of the Atlantic in England, A. D. Stammers reported in 1922 that 2 cc of milk added to a basal diet of rats resulted in normal development. Although the experiments were not strictly comparable to either those of Osborne and Mendel or of Hopkins, they did show a significant change in the growth of rats that had been on an A-deficient diet for more than three months. Eight of ten rats exhibited all the symptoms of xerophthalmia (an eye disease brought on by lack of vitamin A). After receiving 2 cc of milk per rat per day for more than three months on a diet totally deficient in vitamins, the xerophthalmia disappeared and the rats all increased in weight. (54)

At about this time (1922), the conflict over quantitative relations of milk seems to have disappeared from the scientific literature. So far, no satisfactory or plausible explanation of the differences between Hopkins' results and those of Osborne and Mendel has appeared in the history of nutritional biochemistry, and there is a very real possibility that no explanation will ever be found for the simple reason that the need to discover a solution has vanished. (For the moment we will ignore the possible solution by a future historian who will move heaven and earth to obtain an answer to what was once a vexing problem.)

In order to make sense out of the remark that the need to know has vanished, we will turn to some specific aspects of those experiments that could not be reconciled, derive a bit of understanding of the formidable nature of the problem and then retire quietly, having assured the reader that the author has indeed found the "best" of all possible answers.

Hopkins' 1912 paper (55) is, at first glance, superlative, specific in detail, statistically valid and graphically persuasive. A more thorough study reveals glaring omissions, numerous arithmetical errors, a lack of some critical details and a series of graphs so constructed as to be misleading (five different graph scales to represent seven graphs.)

The statistical presentation of the protocols in the Appendix is of enormous value to the reader. That it should be marred by a minimum of 25 arithmetical errors is annoying and makes one wonder who did the checking of data, but, as it turns out, the overall conclusions of the paper are not negated by these errors. (Unless, of course, a reader is a complete skeptic and cynic to boot, believing that such arithmetical flaws are indicative of experimental flaws and hence, the entire paper becomes suspect.)

Hopkins was careful to apply the governing rule of nutrition to his experiments: all the dietaries were more than adequate with respect to their energy content. With rare exceptions the rats used in his studies consumed sufficient food, fats and carbohydrates in particular, to supply them with the calories deemed necessary for growth. The quantities of protein and of mineral salts were also adequate. With all this in mind, attention can now be focused on the specific diets, experimental procedures and results.

The two dietaries employed by Hopkins are listed as follows:

	Pure Casein Mixture (A)	"Protene" Mixture (B)
Protein	22.0%	21.3%
Starch	42.0	42.0
Cane sugar	21.0	21.0
Lard	12.4	12.4
Salts	2.6	3.3

The protein in both foods was casein, the pure casein being a laboratory grade obtained from the Merck Co. while "Protene" was a commercial preparation of casein, not as purified as the Merck's.

The salts were obtained by combustion of the normal food supplied to rats when not under experimental study, namely equal parts of oats and dog-biscuits. The nutrient composition of the dog biscuits or the oats was not given in the paper but it can be assumed that the food was adequate for normal growth and development of rats.

Both of the above diets supplied almost exactly 5 Cal/g of energy. Hopkins reported the food consumption of experimental rats in terms of calories per 100 g of live weight of the animals, and it is a relatively simple task to work from this data and obtain the actual food consumption of the animals.

Rather than reproduce the graphs or protocols exactly as published, the author has selected some of the data in recalculated form in order to bring out the more salient aspects of the experiments as Table I illustrates.

Observations, Anomalies, Comments, Interpretations

The purification of foods with hot alcohol was specified for some of the experiments. Hopkins said that when he used extracted diets he also employed specially purified lard but he did not explain how the fat was purified.

Aside from experiment 7, all the others were of short duration. Furthermore, the rats used in experiment 7 were all ini-

Table I. Summary of Hopkins' 1912 Paper. Effects of Small Milk Supplements on Rat Growth

Exper. No.	Dietary	No. of Rats	Initial Avg. Wt.	Final Avg. Wt.	Gain Avg.	Time Days	Avg. Gain Per Day Per Rat	Avg. Food Intake, Per Rat Per Day
1	A, unext.** +2 cc	6	36.4 g	93.3 g	56.9 g	36	1.58 g	8.2 g
1	A, unext.	6	36.6	38.7	2.1	23*	0.09	4.3
2	A, ext. ÷ 3 cc	8	44.4	74.6	30.2	18	1.68	NA
2	A, ext.	8	44.6	48.1	3.5	18	0.19	NA
3	A, ext. +2 cc	6	41.3	66.8	25.5	19	1.34	6.8
3	A, ext.	6	41.3	45.2	3.9	19	0.20	5.3
5	B, part. ext. + 3cc	8	41.4	92.1	50.7	25	2.03	ca. 6.7
5	B, part. ext.***	8	40.0	60.1	20.1	25	0.80	ca. 5.0
6	B, unext. + 5 cc	6	38.5	123.3	84.8	34	2.49	ca. 8.4
6	B, unext. + 2 cc	6	37.7	112.2	74.5	34	2.19	ca. 7.4
6	B, unext.	6	38.5	85.5	47.0	34	1.47	ca. 6.4
7	B, unext. + 5 cc	4	114.0	219.3	105.3	61	1.73	ca. 13
7	B, unext.	4	113.7	165.0	51.3	61	0.84	ca. 11

* After the 23rd day these rats began to fail. Five were dead by the 31st day.
** Unextr. = unextracted with alcohol; ext. = extracted with alcohol
*** Part ext. means partial extraction of the protene. Hopkins did not explain what he meant by partial extraction.
NA Not Available. No data provided in the protocol.
Milk addendum shown in cc and measured as cc per rat per day.
Aside from the milk fed rats, nearly all others showed some growth during the first 7 to 12 days.

tially much heavier than most of the others in the remaining experiments. In experiment 7, rats receiving milk gained more per day during the first half of the time period than in the second, while those on "protene" alone gained more than twice as much per day in the second half of the experimental period than in the first.

While he was not able to supply equal numbers of male and female rats for these studies, there were sufficient rats of both sexes to warrant Hopkins' statistically derived conclusions that a small milk addendum is a growth promoter at a rate far in excess of what might be expected from the addition of the milk fat, proteins, carbohydrate and salts contained in 2-5 cc of whole milk.

The problems of interpretation are, as mentioned earlier, multi-factored. The sex, age, weight and general health of the rats have to be taken into account as do their pedigrees or genetic makeup. The energy supply of the nutrients must be sufficient and the amino acids in terms of the proteins used must be qualitatively adequate. Temperature, relative humidity, caging techniques and handling are also important factors. Then there are the characteristics of the cows from which milk was obtained. The list can be extended but insofar as he was able, Hopkins appears to have taken many of these factors into account. Most importantly, he used a large number of rats to make his results statistically significant by averaging out individual differences, leading to the conclusion that milk contains "accessory growth substances."

In somewhat later terminology, Hopkins' rats were apparently receiving both fat-soluble A and water-soluble B in their diets. The question is: Where did they get these vitamines in his experiments? A likely source in young rats is the 'store', postulated by Osborne and Mendel, obtained during the suckling period. However, based upon present-day knowledge only fat-soluble A is retained in certain fatty deposits of the body. Young rats, just weaned (as most of Hopkins' rats appear to have been, judging from their weights) would have variable quantities of A in their tissues and an extremely limited source of water-soluble B which is stored poorly in most animals.

Since the premise was that rats could not grow without both A and B (although Hopkins would not have known this in 1912), then any growth whatsoever indicated that some of the essential A and B were present in his most successful diets.

It should also be said again that casein, the complete protein (in terms of its amino acid content), was obtained from milk. Whether highly purified by the Merck Co. or not (as in "protene"), we have noted Hopkins' insistence that only hot alcohol could extract the growth substances completely. Casein, Hopkins believed, could also adsorb trace amounts of these substances. McCollum, Drummond and Osborne and Mendel would all demonstrate that lactose, also obtained from cow's milk, adsorbed variable quantities of growth promoting materials. (Anyone who has left a quantity of milk in a refrigerator is aware that milk 'picks up' odors and flavors of other foods. Most of these adsorbed smells and tastes are found in company with lactose although casein can also adsorb quite well.)

Another potential source of fat-soluble A in these diets was the lard. Hopkins said that it was purified sometimes, but how? And, in the case of unpurified lard, did he use leaf lard from the back of the pig, lard known (later) to be lacking in A or did he use "keg" lard, a mixture of all pig fats, some of which contain fat-soluble A? We simply do not know.

Keeping all this in mind, Table I can be interpreted a different way. Experiments 2 and 3 showed some growth in rats not receiving milk. Their limited success can be attributed to a small store of both A and B with the B depleted most rapidly. There is also the possible availablility of very small amounts of A and B not extracted by Hopkins' method. Experiments 5, 6 and 7 all illustrated that commerical "protene" contained both the A and B factors.

Why, then, were Osborne and Mendel unable to validate Hopkins' results? Why couldn't they obtain significant growth until they added 10-16 cc of whole milk per day per rat to their diets? Were American rats so different from English ones? Were English dairy cattle much more productive in nutrients than American ones?

Careful scrutiny of their publications, draft manuscripts, correspondence, record books, idea and plan books along with marginal notes made on their copies of various journal articles has resulted in an hypothesis that is too long to be elaborated in its entirety. A few bits and pieces will suffice for now. First of all, Osborne and Mendel believed their own experimental results and conclusions to be more correct than others unless the others agreed with them completely (not an unusual phenomenon). Second, they were convinced that Hopkins' experiments were of too short a duration to truly compare his results with theirs in which p. f. m. had been used as a successful supplement for up to 80 days. Third, Osborne and Mendel were convinced of the purity of the diets they used. They prepared most of the proteins in their own labs in addition to the p. f. m. It must have been with some reluctance that they finally admitted that commerical lactose or even some of their 'home-made' lactose carried growth promoters.

The matter of experimental duration vexed them greatly. They had maintained rats without significant growth for months on diets which, as far as they could see, were equivalent to those of Hopkins on which his rats did not survive very long.

With the United States eventually entering the war in Europe in 1917 came the impetus to develop better foods that had long shelf lives. By this date the vitamines were firmly entrenched in nutrition and it is not surprising that great emphasis was placed upon the production of milk and other dairy products since everybody knew that milk was good for you.

Osborne and Mendel took up the issue of the vitamine content of milk once again in 1918 and attempted to reproduce Hopkins' results. At this date they were certain that the 'missing' vitamine was water-soluble B so they prepared diets to include about 9% butterfat to supply the necessary A factor. In their publication of 1918 it is evident that the diets were not exact replicas of Hopkins' dietaries; nevertheless, the critical nutrients were available to the rats in ample quantity, or so it appeared.

Again, to simplify matters, significant data are presented in tabular form (Table II) while the diets are described below. Osborne and Mendel did not present the large bodies of statistical information as Hopkins did so rat weights and times of experiment must be interpolated from the graphs of rat growth.

Because the authors complicated the graphs by making many changes in the diets, the growth curves are not simple. Therefore, in Table II will be found only those portions of the experiments in which milk was given as an addendum in approximately the same quantities as in Hopkins' experiments. Two other points need to be mentioned. All the rats used by Osborne and Mendel were males and none of the foods were extracted with alcohol.

Note that the salts in diet 6 were prepared in a fashion similar to the technique used by Hopkins: incineration of equal parts of oats and dogbread. Also note that the percentages in Diet 7 add up to 105. The error was probably in the figure for starch.

Osborne and Mendel were correct in believing that if cow's milk provided the necessary vitamines in small quantities such as Hopkins had found to be effective, their experiments ought to have shown this unequivocally even if the methods and basal diets were somewhat different. Yet, they could not obtain growth in rats as Hopkins did. Why not? A careful study of Tables I and II reveals that Osborne and Mendel's rats did not eat as much total food per day as did Hopkins' rats. Osborne and Mendel were aware of this phenomenon as was Hopkins, presumably.

But this only adds to the mystery. Examination of the growth curves of the 1922 paper in particular shows that rats being given a milk supplement on a B-deficient diet grew poorly if at all. Yet, as Hopkins had found years earlier and as Osborne and Mendel had discovered on their own, the addition of small amounts of yeast or even smaller quantities of a yeast extract to the diets of such rats coincided with an increased food intake and a related growth spurt.

At this point in 1922 we shall take leave of the conflicting, annoying and confusing results of such animal feeding experiments. Let it suffice to say that if the average daily gain in weight is used as a criterion, then Hopkins' rats did very well indeed. But, by way of comparison, today's laboratory rats (males) will routinely gain from 5.0 to 5.7 g per day during growth from a weight of about 45 g to 220 g while consuming 9 to 21 g of food per day. (56)

Table II. Summary of Osborne and Mendel Papers, 1918-1922
Effects of Small Milk Supplements on Rat Growth

Paper	Dietary	No. of Rats	Initial Avg. Wt.	Final Avg. Wt.	Gain Avg.	Time Days	Avg. Gain Per Day Per Rat	Avg. Food Intake, Per Rat Per Day
1918	Diet 1 + 2 cc	4	72.0 g	78.5 g	6.5g	14	0.46 g	NA
1918	Diet 2 + 2 cc	3	60.3	85.3	25.0	40	0.62	NA
1918	Diet 2 + 2 cc	1	78.0	98.0	20.0	47	0.43	NA
1918	Diet 3 + 2 cc	3	74.7	90.3	15.6	21	0.74	NA
1918	Diet 4 + 4 cc	3	Two of the rats declined immediatley; the other showed a small gain for a few days before declining.					
1920	Diet 5 + 2 cc	2	60.5	64.0	3.5	35	0.10	3.3 g, 14 d*
1920	Diet 5 + 2 cc	1	65.0	70.0	5.0	42	0.12	5.7 g, 14 d*
1920	Diet 5 + 5 cc	2	64.0	72.5	8.5	18	0.47	3.7 g, 14 d*
1920	Diet 5 + 5 cc	1	70.0	103.0	33.0	49	0.67	4.1 g, 35 d*
1920	Diet 5 + 5 cc	3	66.0	102.7	36.7	38	0.95	ca. 5 g, 14 d*
1920	Diet 6 + 2 cc for 21 d then 5 cc for 28 d	3	Negligible growth with either quantity of milk.					
1922	Diet 7 + 2 cc	6	40.8	53.7	12.9	10	1.29	ca. 3.6 g
1922	Diet 7 + 2 cc	6	40.8	59.0	18.2	20	0.91	ca. 3.6 g**

* The authors provided intake data for only specific periods. The values given here are the intakes per day per rat over the time given, in days, d.

** The 1922 experiment used the same 6 rats. For comparison purposes the data has been separated into the first 10 days and the total elapsed time of 20 days.

Milk addendum shown in cc and measured as cc per rat per day.

Addendum to Table II
Composition of Diets

Diet 1	
Casein	18.0%
Salts	4.5
Starch	50.5
Butterfat	9.0
Lard	18.0

Diet 2	
Edestin	18.0%
Salts	0.0-4.5
P. f. m.	28.0-0.0
Starch	25-48.5
Butterfat	9.0
Lard	20.0

Diet 3	
Casein or Edestin	18.0%
P. f. m.	0.0-22.4
Salts	4.5-1.0
Starch	47.7-31.6
Butterfat	9.0
Lard	18-20.0

Diet 4	
Casein	18.0%
Milk Pwdr.	0.0-48.0
Salts	4.5-1.0
Starch	50-18.0
Butterfat	9.0
Lard	18.0

Diet 5	
Casein	18%
Salts	4
Starch	49
Butterfat	9
Lard	20

Diet 6	
Casein	18.0%
Salts	4.5
Starch	50.5
Butterfat	9.0
Lard	18.0

Diet 7	
Casein	18%
Salts	4
Starch	54
Butterfat	9
Lard	20

Nevertheless, it is remarkable that the pioneers of the vitamin hypothesis were able to do so much when for twenty years or more they had no clear conception of the chemical entities with which they were dealing. Contradictory or not, those animal feeding experiments opened up new vistas in nutrition and in all the intervening years since the vitamin hypothesis came into being there has not been any single 'breakthrough' in nutrition of comparable magnitude and of such pervasive influence.

The Author Speaks but Hopkins has the Last Word, Almost

In following the controversy over vitamins in milk, I have been aided by having many of Osborne and Mendel's records, ideas, manuscripts and correspondence at hand, thanks to the historical tradition at the Connecticut Agricultural Experiment Station. Perhaps the Hopkins' papers (such as exist) would enable me to solve the 'quantitative relations' issue but, as stated earlier, I do not think it will ever be solved.

In the early 1920's enough reliable information about vitamins had been obtained to render the problem unimportant. Many different foods were found to contain one or more vitamins and the role of milk in the diets of humans, at least as a supplier of vitamin B, fell into relative unimportance.

Although dairy interests continue to advertise milk as the 'perfect' or 'near perfect' food, the fact remains that cows' milk is not a very good source of B vitamins in terms of percent composition by weight. After all, milk solids constitute only 12-13% of fluid milk and that means a rather large intake ratio of water to solids (about 7/1), not exactly ideal for obtaining trace nutrients such as the B vitamins.

As far as Hopkins' Nobel Prize is concerned, he deserved it even though his experimental "proofs" could not be replicated in a consistent fashion. He deserved the prize for one reason above all others: persistence in promulgating an idea. At a time when the energy doctrine controlled nutrition, Hopkins was questioning its absolute validity. And while he was not the first to do so, he was the only one of the early questioners to proclaim his beliefs in print and to persist in a search for the elusive 'accessory growth factors', a search that required not only experimental work of his own but also a commitment to urge and train others to follow in his footsteps.

Earlier, I mentioned that Casimir Funk had raised the issue of priority in the discovery of the vitamins. Hopkins did not enter into the debate at the time but he did devote a significant part of his Nobel Address to Funk and his contributions to the vitamin hypothesis. With respect to the other principals involved in the early episodes of vitamin research, it is appropriate to see what Hopkins said about them in Stockholm on that December day in 1929. His comments afford one more illustration of both the clarity and confusion that dominated his 1912 paper.

In order to make sense of this vague statement, let me remind the reader that in 1917 McCollum had moved to John Hopkins in Baltimore. Osborne spent his entire professional career at the Connecticut Agricultural Experiment Station while Mendel had been a faculty member of the Sheffield School of Physiological Chemistry at Yale University. Their joint publications always recognized the two institutions but somehow, over the years, it was Yale that appeared to be the spawning ground of their nutritional researches and discoveries.

At the close of his Nobel Address, Hopkins said this about McCollum and Osborne and Mendel:

> So prominent indeed was the American work at this time [World War I] and so large a proportion did it form of the total output from 1912 to near the end of the war that, if I wished to claim that my own publications exerted any real and effective influence in starting a new movement in the study of dietetics, I should have to convince myself that they helped to direct the thoughts of the Harvard and Baltimore investigators.(57)

Well, Hopkins may have been confused in his Yale/Harvard geography and his 1912 paper showed that he had problems with simple addition and division but he certainly knew how to promote multiplication in rats.

When all is said and done I can respond to the title of this paper, Will Milk Make Them Grow? by saying yes, sometimes; no, frequently; and maybe, under certain conditions. Such a diffuse and qualified response is illustrative of the difficulties inherent in obtaining consistent results with animal feeding experiments during the first two decades of the twentieth century.

Literature Cited

All correspondence items are located in the Thomas B. Osborne Files, Department of Biochemistry, The Connecticut Agricultural Experiment Station, New Haven, Connecticut.

1. Funk, C. J. State Med. 1912, 20, 341-368.
2. Hopkins, F. G. J. Physiol. 1912, 44, 425-460.
3. McCollum, E. V.; Davis, M. J. Biol. Chem. 1913, 15, 167-175.
4. Osborne, T. B.; Mendel, L. B. J. Biol. Chem. 1913, 15, 311-326.
5. Osborne, T. B.; Mendel, L. B. "Feeding Experiments with Isolated Food-Substances"; Carnegie Institution Publication 156, Washington, D. C. 1911; Part I, 1-53; Part II, 55-138.
6. McCollum, E. V. "A History of Nutrition"; Houghton Mifflin, Boston, 1957, Chapters 1-14, 18, 20, 21.
7. Becker, S. L. "The Emergence of a Trace Nutrient Concept through Animal Feeding Experiments"; Unpublished Doctoral Dissertation, University of Wisconsin, 1968, Chapters 1-3.
8. Funk, C. Science. Apr. 30, 1926, 63, 455-456.
9. van Leersum, E. C. Science. Oct. 8, 1926, 64, 357-358.
10. Needham, J.; Baldwin, E. (eds.) "Hopkins and Biochemistry": W. Heffer and Sons, Ltd., Cambridge, 1949, 198-199.
11. Hopkins, F. G. The Analyst. 1906, 31, 395-396.
12. Hopkins, F. G.; Neville, A. Biochem. J. 1913, 7, 97-99.
13. Hopkins, F. G. Biochem. J. 1920, 14, 721-724.
14. Becker, S. L. "Butter Makes them Grow"; Bulletin 767 of Connecticut Agricultural Experiment Station. 1977, 1-21.
15. See citation 3.
16. Funk, C. J. Physiol. 1913, 46, 173-179.
17. Hopkins, F. G. J. Indust. Engr. Chem. 1922, 14, 67-68.
18. See citation 10, 197.
19. Osborne, T. B.; Mendel, L. B. J. Biol. Chem. 1912, 12, 81-89.
20. McCollum, E. V.; Kennedy, C. J. Biol. Chem. 1916, 24, 491-502.
21. Drummond, J. C. Biochem. J. 1920, 14, 660.
22. See citation 5.
23. Letter, Hopkins, F. G. to Osborne, T. B., August 4, 1912.

24. Letter, Mendel, L. B. to Osborne, T. B., August 17, 1912.
25. Letter, Mendel, L. B. to Osborne, T. B., August 30, 1912.
26. See citation 5, 80-85.
27. Osborne, T. B.; Mendel, L. B. Proceed. Soc. Exper. Biol. Med. April 17, 1912, 9, 72-73.
28. Letter, Osborne, T. B. to Mendel, L. B., July 18, 1912.
29. See citation 12, 97.
30. Ibid., 99.
31. Letter, Osborne, T. B.; Mendel, L. B. to Hopkins, F. G., Feburary 28, 1913.
32. See citation 19, 83-84.
33. Letter, Hopkins, F. G. to Osborne, T. B.; Mendel, L. B., March 14, 1913.
34. Ibid., 1.
35. Ibid., 6.
36. Letter, Osborne, T. B.; Mendel, L. B. to Hopkins, F. G., undated draft manuscript, but probably late March to mid-April, 1913.
37. Ibid., 9.
38. McCollum, E. V.; Davis, M. J. Biol. Chem. 1915, 23, 184.
39. McCollum, E. V.; Davis, M. J. Biol. Chem. 1915, 23, 234.
40. Drummond, J. C. Biochem. J. 1916, 10, 89-102.
41. Osborne, T. B.; Mendel, L. B. J. Biol. Chem. 1917, 31, 149-163.
42. Osborne, T. B.; Mendel, L. B. J. Biol. Chem. 1919, 37, 187-200.
43. Osborne, T. B.; Mendel, L. B. J. Biol. Chem. 1919, 37, 572.
44. Osborne, T. B.; Mendel, L. B. J. Biol. Chem. 1918, 34, 539.
45. Ibid., 543-4.
46. Osborne, T. B.; Mendel, L. B. J. Biol. Chem. 1920, 41, 515-523.
47. Osborne, T. B.; Mendel, L. B. Biochem. J. 1922, 16, 363-367.
48. Letter, Osborne, T. B.; Mendel, L. B. to Hopkins, F. G., December 8, 1919, 2.
49. Ibid.
50. Letter, Hopkins, F. G. to Osborne, T. B.; Mendel, L. B., February 2, 1920, 2-3.
51. Letter, Osborne, T. B; Mendel, L. B. to Hopkins, F. G., February 26, 1920, 1-2.
52. Letter, Mendel, L. B. to Hopkins, F. G., July 1, 1921.
53. Dutcher, R. A. J. Indust. Engr. Chem. 1921, 13, 1103.
54. Harrow, B. Chem. Abst. 1923, 17, 411. Original article: Stammers, A. D. Biochem. J. 1922, 16, 659-667.
55. See citation 2.
56. Altman, P. L.; Dittner, D. S. (eds.) "Biology Data Book"; Fed. Amer. Soc. Exper. Biol., Bethesday, MD, 1974, 3, 1457.
57. See citation 10, 200.

RECEIVED March 8, 1983

5

Charles Holmes Herty

And the Effort to Establish an Institute for Drug Research in Post World War I America

JOHN PARASCANDOLA

University of Wisconsin, School of Pharmacy and Department of the History of Science, Madison, WI 53706

This paper discusses the attempt to establish an institute for drug research in the United States in the period 1918-1930, a campaign that did not achieve its original objective but did play a significant role in the creation of the National Institute of Health. The First World War had clearly demonstrated American dependence upon German chemicals such as synthetic dyes and drugs, and the effort to establish an institute for drug research must be viewed in the context of the concern for achieving American independence in the chemical industry. The story of this effort, which was spearheaded by chemist Charles Holmes Herty, helps to illuminate certain broader trends in the chemical and pharmaceutical fields at the time.

"A few days ago we asked a well-known organic chemist, one who has been particularly successful in working out the methods for the manufacture of certain much-needed coal-tar medicinals, 'Suppose during your researches you made some new compound which you believed would prove more efficacious against certain diseases than any of the known compounds whose details of manufacture you have solved, where would you turn to have it tested thoroughly.' He replied, 'I don't know'" (1).

The conversation described above took place at a subway station as the participants were returning from a baseball game at the Polo Grounds in New York in the late summer of 1918. The organic chemist referred to was Professor J. R. Bailey of the University of Texas (2). The narrator, who posed the question, was Dr. Charles Holmes Herty, Editor of the American Chemical Society's Journal of Industrial and Engineering Chemistry. Herty was reporting on this discussion in an editorial in the September 1918 issue of the Journal, an editorial that was to start a campaign to establish a national institute for drug research. Although the plan never came to fruition, it is worthwhile

0097–6156/83/0228–0085 $06.00/0

examining the history of this effort for several reasons. It was through the committee that was established for this purpose, for example, that Herty came to play a key role in the creation of the National Institute of Health. The story of the effort to establish such an institute also helps to illuminate certain broader developments occurring within the chemical and pharmaceutical fields in the decade following the end of World War I.

When the First World War began, the dependence of the United States upon foreign countries, especially Germany, for its supply of synthetic organic chemicals such as dyes and drugs soon became obvious. As one representative of a major American pharmaceutical firm later commented: "Before the Great War, this country cut a small, indeed, an almost contemptible figure in the manufacture of synthetic chemicals of all kinds, and particularly in the production of synthetic medicinals" (3). The United States was not alone in relying on Germany for such chemicals. About ninety percent of the dyes and almost all of the coal-tar medicinals used by the British in the period immediately preceeding the war were imported from Germany (4). The German chemical industry dominated the synthetic organic chemical field.

The synthetic drug industry in Germany had largely grown out of its strong coal-tar dye industry in the late nineteenth century. Seeking uses for the by-products of dye manufacture, these firms, which pioneered in the development of industrial research laboratories, set their chemists to the task of developing products of pharmacological interest. A host of synthetic analgesics, antipyretics, anesthetics and hypnotics, such as Phenacetin, Aspirin, Novacaine and Veronal, were introduced onto the market by German firms. In 1910, the first of the successful synthetic chemotherapeutic agents, Salvarsan, was introduced by Paul Ehrlich, and manufactured by the Höchst Chemical Works. Used in the treatment of syphilis and certain trypanosome infections, it quickly became one of the most important of the patented German synthetic drugs (5, 6, 7).

As the War in Europe progressed, it became increasingly difficult for the United States to obtain the needed synthetic drugs and other chemicals from Germany. After the United States entered the conflict in 1917, the Trading-with-the-Enemy Act was passed, giving the Federal Trade Commission Authority to license patents owned by enemy aliens. Selected American firms were licensed, for example, to manufacture patented synthetic drugs such as Salvarsan. The War thus provided a stimulus to the American pharmaceutical manufacturing industry, and prompted a number of American companies to enter the synthetic drug market (8). After the war ended, the major American pharmaceutical firms continued to produce synthetic drugs and made every effort to retain their share of the market. Concern was expressed about the Germans regaining their former dominance in the field of synthetic chemicals, and tariff legislation was passed to protect the infant American synthetic dye and drug industries in the domestic market (9, 10).

In the postwar years a spirit of "chemical boosterism," to use a term coined in a recent historical study, emerged in the United States. A group of chemists, chemical journalists and science popularizers acted as propagandists for the chemical profession and industry. Their message, delivered with conviction to the public through a variety of popular publications, was that chemistry was vital to the national defense and to economic progress. One of the more active of these chemical boosters was Charles Holmes Herty, and his efforts to establish a national institute for drug research must be viewed within this context (11).

Born in Georgia in 1867, Herty was the son of a pharmacist and originally planned to enter his father's profession. After graduating from the University of Georgia with a Ph.B. in 1886, however, he decided to undertake advanced study in chemistry under Ira Remsen at The Johns Hopkins University. Herty received his Ph.D. from Hopkins in 1890, and, after a year at the Georgia State Experiment Station, joined the faculty of the University of Georgia, where he remained for the next decade. During that period he took a leave of absence to pursue postdoctoral work at the universities of Berlin and Zurich. While in Berlin, he attended the lectures of the noted synthetic dye chemist Otto Witt.

It was Witt's criticism of the method of collecting turpentine from pine trees in the American South that stimulate Herty's interest in this subject. Upon returning to the United States in 1901, he turned his attention to this problem, and he soon developed his cup and gutter system, which replaced the old "boxing" method. Herty resigned his university post and worked with the United States Bureau of Forestry (1902-1904) and later the Chattanooga Pottery Company (1904-1905) in following up on his invention, which made him financially secure.

He returned to teaching in 1905 when he accepted a call to head the Department of Chemistry at the University of North Carolina. Herty retained a strong interest in the practical applications of science, and served as Dean of the School of Applied Science at North Carolina from 1908 to 1911. He achieved sufficient national prominence to be elected President of the American Chemical Society in 1915 and again in 1916 (12, 13, 14).

In his presidential address of 1915, Herty stressed the issues that were to occupy his concern for much of the rest of his career, calling for greater cooperation between universities and industry and a strengthening of the American chemical industry (15). When he became editor of the American Chemical Society's *Journal of Industrial and Engineering Chemistry* in 1917, he was provided with a convenient forum for his preaching of the chemical gospel. During the five years in which he edited this journal, Herty energetically promoted the development of American chemistry. He campaigned for the establishment of an American synthetic organic chemical industry, for protective tariffs to enable it to survive and flourish, for the allocation of chemists to war work, and for the creation of the Chemical Warfare Service (16).

It is thus not surprising that Herty took an interest in the question of the development and testing of chemical agents for the treatment of disease. After the above-mentioned 1918 conversation with Bailey stimulated his thinking on the subject, Herty decided to devote an editorial to discussing the need for better pharmacological laboratory facilities for organic chemists to test potential new medicinal chemicals. He received the encouragement of Simon Flexner and P. A. Levene of the Rockefeller Institute for Medical Research, whom he consulted about the idea (17).

In his editorial, entitled "War Chemistry and the Alleviation of Suffering," Herty pointed out that in the area of medicinal chemistry the chemist cannot work alone. He must rely on the pharmacologist and the physiologist to determine the therapeutic potential of a product. Cooperation between the chemist and the biologist was thus essential. Herty complained that universities often lacked the funds and the necessary spirit of cooperation to undertake such studies, that most manufacturing establishments had inadequate facilities for these purposes, and that government laboratories suffered from insufficient appropriations for research. Certain privately-endowed institutions, most notably the Rockefeller Institute, provided the appropriate environment for such cooperative research, but there were few such institutions and their capacity for work was necessarily limited.

Wartime research, e.g. on gases and explosives, had developed a corps of organic chemists whose talent could be applied to the relief of suffering if one could devise an appropriate mechanism. What was needed for this application, Herty argued, was an institution, somewhat analagous to the Mellon Institute, with adequate provision for laboratory tests of all kinds and to which, through the establishment of fellowships, manufacturing firms could send personnel for working out specific problems. He saw this proposed institution cooperating with university laboratories of organic chemistry and with hospitals in carrying out its work. Such an institute would also stimulate the development of more adequate research facilities within the pharmaceutical industry itself (1). The specific idea for such an institute was apparently suggested to Herty by Levene (18, 19).

Herty's editorial closed by soliciting comments from the readership. He received at least a few letters, but recognized the need to be more aggressive in promoting discussion of the subject. Herty thus arranged for the November 8, 1918, meeting of the New York Section of the American Chemical Society, which he served as Chairman, to be devoted to the topic (20). Seven prominent individuals spoke to the need for an institute for cooperative research as an aid to the American drug industry.

These included Johns Hopkins pharmacologist John J. Abel (by letter only, as he could not attend), Rockefeller Institute biochemist P. A. Levene, Chief of the U.S.D.A. Bureau of Chemistry Carl Alsberg, Wisconsin pharmacologist Arthur Loevenhart (engaged at the time in chemical warfare work), Acting Director of the Mellon Institute for Industrial Research E. R. Weidlein, and two industry representatives, Frank Eldred of Eli Lilly and Company and D. W. Jayne of the Barrett Company. The addresses were published in the December 1918 issue of the Journal of Industrial and Engineering Chemistry, and were also reprinted and circulated to numerous individuals whose views were solicited (21).

A significant number of letters were received in response to this request and were published in the Journal over the next few months (22). Basically all of the correspondents favored the idea of an institute for drug research, although they differed somewhat in their vision of how it might operate. In the meantime, the New York Section adopted a resolution to refer the matter to the Advisory Committee of the American Chemical Society, urging that the proposed institute be undertaken under the auspices of the Society. The Advisory Committee authorized President William Nichols to appoint a committee to report on the endowment that would be needed to support such an institute and to outline the policies under which it should be operated. The Committee on the Institute for Drug Research was appointed in February 1919 with Herty, to nobody's surprise, as Chairman. Three of the participants of the New York symposium (Abel, Levene, and Eldred) were also on the Committee, along with Harvard pharmacologist Reid Hunt, Yale chemist Treat Johnson, Director of the Mellon Institute Raymond Bacon, and chemist F. O. Taylor of Parke, Davis and Company (23, 24).

The Committee moved quickly into action, holding a meeting in New York on February 22 and developing a draft report by early March. The report called for the creation of an Institute for Drug Research with three major divisions: chemistry, pharmacology and experimental biology. It was estimated that an endowment of about ten million dollars would be needed to fund an institute of the size and scope envisioned. The total proposed staff (including laboratory assistants and office personnel) exceeded one hundred. The permanent staff would devote itself to pure research, but manufacturers would be offered facilities for the solution of their specific problems through the creation of industrial fellowships. It was also suggested that the institute be controlled by two boards, a financial board of directors, composed of able financiers, and a scientific board of directors, composed of eminent scientists.

The report concluded that the financing of the institute was too great an undertaking for the American Chemical Society or the drug manufacturers, and that it was undesirable to seek state or federal aid for the kind of research contemplated. Instead, the Committee recommended that the institute be privately endowed, the funds to be provided by "a big-hearted American or Americans" (25, 26, 27). The idea of finding a rich benefactor to support the institute had already been expressed by Herty in his original editorial. Even before the Committee had been appointed, Abel voiced a similar hope when he wrote to Herty that what was needed was to "fire the imagination of some old millionaire who is about to join his ancestors and wishes to put his millions to the best possible use" (28). This hope of finding a "sugar daddy" to bankroll the project, however, was destined never to be realized.

Apparently not all of the members of the Committee were satisfied with the draft report, for Herty indicated that several had made important suggestions for changes. He therefore called another meeting of the group at the American Chemical Society meeting in Buffalo on April 9. The meeting was to follow a special symposium, sponsored by the Pharmaceutical Division, on "The Possibilities of Drug Research" (29, 30). Frank Taylor, a member of Herty's Committee, was Chairman of the Division, and felt that such a topic was timely in light of the recent discussions concerning the proposed institute for drug research (31).

The Committee was not to actually develop the final version of its report, however, until more than two years later, and when it did eventually appear the scope of the report and of the proposed institute had been considerably expanded. One reason for the delay was that Herty found himself becoming occupied with more pressing matters in 1919, and no doubt had less time to devote to the work of the Committee. As Secretary of the Dye Advisory Committee of the War Trade Board, Herty became involved in the negotiations concerning the purchase of German dyestuffs through the Allied Reparations Commission. He was chosen to go abroad to arrange for the purchase and shipment of dyes not available from domestic manufacturers (32). Herty also became a participant in the post-war struggle for tariff legislation to protect the American dye industry (33, 34, 35).

Criticisms of the proposed institute began to appear in 1919, and perhaps also served to distract Herty to some extent. The criticisms would appear to have stemmed at least in part from interprofessional rivalries. In April the *Chicago Chemical Bulletin*, a publication of the Chicago Section of the American Chemical Society, published a critique of the proposed institute that perplexed and concerned Herty. Although unsigned, the article was written by Dr. Paul Nicholas Leech, a chemist on the staff of the American Medical Association (36). While granting the merits of an institute for drug research, Leech chastized the proponents of the plan for "keeping the Institute rigidly under the control of the American Chemical Society." The close cooperation of physicians would be needed for the clinical trials essential in the testing of any new drug, yet organized medicine was largely disregarded, Leech complained, in the meetings and published opinions concerning the proposed institute. No official representative of the American Medical Association was asked to participate in these proceedings, in spite of the fact that the Association's Council on Pharmacy and Chemistry had been concerned with the problem of evaluation of drugs since it was founded in 1905. Surely the Council had a better grasp of the synthetic drug situation, Leech continued, than the Mellon Institute or even the Rockefeller Institute, both of which were represented.

Organized pharmacy had also been consulted only "in some slight measure." Leech did not favor a research center controlled by any one group or element, and suggested that a cooperative venture involving the American Medical Association and the American Pharmaceutical Association, as well as the American Chemical Society, would be more successful.

There were two other factors that seem to have also played a role in Leech's attack on the proposed institute. Working in the A.M.A.'s Chemical Laboratory, he was involved in the work of evaluating drugs carried out under the auspices of the Council on Pharmacy and Chemistry, and this experience had obviously made him wary of the pharmaceutical and chemical industries. His article in the Chicago Chemical Bulletin suggested that only a relatively few of the synthetic drugs from Germany were therapeutically valuable. The demand for many others, according to Leech, had been artificially stimulated by "clever advertising." He did not wish to see American chemists flooding the market with "semi-scientific medicaments" (37). In an article on "American-Made Synthetic Drugs" in the Journal of the American Medical Association later that same year, Leech elaborated further on this potential hazard of the institute for drug research: "In view of the agitation to found an institution for cooperative research as an aid to the American drug industry under the auspices of the American Chemical Society, it will be well for the medical profession to be on its guard against too enthusiastic propaganda on the part of those engaged in the laudable enterprise of promoting American chemical industry. Unless it is, it may be inflicted in the future, as in the past, with a large number of drugs that are either useless, harmful or unessential modifications of well-known pharmaceuticals" (38).

The American Chemical Society was closely identified with the chemical industry. Even John Abel, a member of Herty's Committee and a staunch supporter of the proposed institute, expressed some concern that the Society was largely controlled by commercial interests (30). Abel's own professional organization, the American Society for Pharmacology and Experimental Therapeutics, founded in 1908 largely at his initiative, did not even accept industry pharmacologists into its membership (39). On the other hand, industry supporters of the project tended to favor keeping the control of the institute in the hands of the A.C.S. (30, 40). One industry representative wrote to Herty that he felt that pharmaceutical manufacturers would "look with distrust and anxiety upon any arrangement which would allow official appointees of the A.M.A. to play too prominent a part in the organization and government of this proposed new institution" (40). The relationship between the American Medical Association and the pharmaceutical industry was not a particularly harmonious one at the time. In 1915, a committee of the A.M.A. Board of Trustees had expressed the need for institutes for pharmacological research in this country, but warned against "commercialism" in this field, concluding that: "It is only from laboratories free from any relations with manufacturers that real advances can be expected" (41).

Finally, Leech may also have tended to see the proposed institute as a project of the "Eastern establishment." He noted that the principal agitators for the project appeared to be influential members of the New York section and referred to the need to remind the "Eastern chemists" that they could not disregard physicians (37). University of Chicago chemist Julius Stieglitz, who was sympathetic to Herty's plan, felt that the approval of the publication of Leech's critique by the editorial board of the Chicago Chemical Bulletin was an indication of the feeling on the part of many "Western" chemists that chemical activities were too centralized in the East (42). A.C.S. Secretary Charles Parsons also saw sectionalism as a factor in the Bulletin's editorial stand (43). It may have been in an effort to counter this criticism that Stieglitz was eventually added to the Herty Committee.

Criticism of the proposed institute also came from another quarter, namely from the ranks of organized pharmacy. Already in December of 1918, Edward Kremers of the University of Wisconsin School of Pharmacy, one of the nation's leading pharmaceutical educators and a researcher in plant chemistry, wrote to Frank Eldred of Eli Lilly and Company about the institute for drug research. Kremers complained: "But why should American pharmaceutical manufacturers support an institution fostered by the American Chemical Society when pharmaceutical institutions are in the greatest need of all the financial support in sight. I trust that our pharmaceutical manufacturers will prove true to their own calling first" (44).

Kremers also asked whether the American Pharmaceutical Association was going to take a back seat to the A.C.S. again. He apparently wrote to several other manufacturers voicing a similar message (45). A number of other pharmacy leaders joined Kremers in expressing opposition to the proposed plan. One of the most vocal of these critics was H. V. Arny of the New York College of Pharmacy. In an editorial in the December 1918 issue of the Journal of the American Pharmaceutical Association, Arny called for involvement of the A.Ph.A. in the establishment of the proposed institute (46). At the May 1919 meeting of the Philadelphia Section of the American Chemical Society, Arny was more critical of the proposal. He expressed concern lest "commercial influences creep in and ruin the fair edifice about to be erected." He also attacked the proposers for trying to make the institute primarily a creature of the American Chemical Society, and argued instead for the project to be conducted under the joint auspices of the A.C.S., the A.Ph.A., and such national medical associations as might be decided upon. Arny envisioned the institute including a department of pharmacy, as well as departments of chemistry, pharmacology and practical therapeutics (47).

The American Pharmaceutical Association also had hopes of developing a research endowment of its own at this time, and some of its leaders may have seen the proposed institute as a rival for funds. A.Ph.A. President George Beringer had called for the creation of an endowment fund for pharmaceutical research in his presidential address in 1914 (48). In August of 1917, the A.Ph.A. Council established a research fund (49), and shortly thereafter Arny was appointed Chairman of a newly-created research committee (50). Although it was not until early 1920 that the A.Ph.A. Executive Committee formally proposed the formation of a research endowment, with plans calling for the solicitation of funds from industry and other sources (51), it is likely that the idea of soliciting contributions for an endowment fund was already in the minds of some when the research fund and committee was established.

Perhaps more important than the potential competition for funds, however, was a concern on the part of some pharmacy leaders that many scientists failed to acknowledge pharmacy as a branch of science. Pharmacy's pride had just been dealt a severe blow during the First World War when the armed forces refused to recognize it as a profession. Whereas physicians, dentists, engineers and other scientifically-trained personnel were almost always commissioned, pharmacists served in all sorts of capacities outside their field and were almost never commissioned (52, 53). The initiative by the A.C.S. in the area of drug research, without the consultation of organized pharmacy, was seen by some as further evidence of the lack of respect for pharmacy and the A.Ph.A. Arny complained that to certain chemists pharmacy was considered to be merely the running of retail drug stores, and he reminded these individuals that the A.Ph.A. was older than the A.C.S. and that the pages of its Proceedings and Journal were filled with examples of pharmaceutical research (47). George Beal, although himself a Professor of Chemistry at the University of Illinois, also lamented the tendency of chemists to belittle the claims of pharmacy to a place among the scientific professions. Beal, the son of noted pharmacist J. H. Beal, had received his early training in pharmacy and remained sympathetic to and active in the field. He viewed Herty's proposal as implying that the A.C.S. was the only scientific society capable of dealing with problems of drug research, and like Arny, he came to the defense of the A.Ph.A. (54).

One final critic of the Herty proposal should be mentioned, namely Francis E. Stewart, Director of the Scientific Department of the H. K. Mulford Company. Stewart had been trained in both pharmacy and medicine, and had worked in both academia and industry. His objections to the plan centered largely around his identification of the A.C.S. with the chemical industry and with the patenting of medicinal chemicals. Although himself an employee of a pharmaceutical firm, Stewart had long campaigned for a reform in American patent laws in order to prohibit the patenting of drug products (55, 56). He had no objection to manufacturers protecting their interests by patenting processes and machinery, but felt that product patents actually discouraged innovation and competition and were not in the public interest. Stewart also objected to what he believed were abuses in advertising and trademarks on the part of certain commercial interests, citing the German chemical houses as examples. He raised the question as to whether the A.C.S.-associated plan might not serve mainly to promote the commercial interests of the manufacturers of patented chemical products (57).

Stewart suggested that a drug research institute would be a valuable asset if it were placed on an altruistic basis and not exploited for commercial purposes. Like Leech and Arny, he saw a need to bring medicine and pharmacy into the plan. He also argued for an institute that would evaluate all types of materia medica products, not just synthetic drugs (57, 58). In fact, Stewart had for years argued the need for a bureau or agency to evaluate the therapeutic value of new remedies (59). Although he praised the work of the A.M.A.'s Council on Pharmacy and Drugs, he did not feel that it completely fulfilled this purpose. One of the functions of the American Pharmacologic Society, which he founded in 1906, was to carry out investigations, with the cooperation of manufacturers, on the therapeutic and physiological action of drugs, but this plan never got very far in practice (57, 58).

Stewart's firm was heavily involved in the production of biologicals such as the diphtheria antitoxin, and this gave him another reason for criticizing Herty's emphasis on synthetic chemicals. In a memo to Mulford President Milton Campbell regarding the Herty plan, Stewart argued that the proposed institute would favor the use of patented chemical drugs over traditional pharmacopeial remedies and biological products, thus hurting the company's business (60).

Returning to the work of the Committee itself, its efforts were given a renewed stimulus in 1920 when Francis Garvan, President of the Chemical Foundation, became interested in the proposed institute. The Chemical Foundation had been created in 1919 to handle the licensing of foreign chemical patents abrogated during the war and to promote chemical industry and research with the profits so derived (61). Garvan and the Foundation played a central role in the "chemical boosterism" referred to earlier (11). Having lost a child himself to infectious disease, Garvan was also a strong supporter of medical research (62).

Garvan offered the Committee $50,000 to be used to bring the plans for the institute into more definite shape, and perhaps toward initiation of certain phases of the work (63). On October 11, 1920, he met with the Committee in New York to discuss the project. Garvan felt that the Committee should take the time to develop "a thoroughly systematic survey of the country's resources, present efforts, lines of coordination, and suggestions for more concentrated effort" in the form of a definite report that would outline a program for the application of chemistry to medicine. He pledged the support of the Chemical Foundation in the development of such a report (64).

At this meeting, it was decided to appoint a subcommittee to prepare a draft of "a report in popular language of the best means of promoting chemical science for the prevention and cure of disease" (64). As eventually appointed by Herty, the subcommittee consisted of Herty himself as ex-officio Chairman, Stieglitz, Johnson, Hunt and Eldred (65). Presumably under the influence of Garvan and the Chemical Foundation, the scope of the research institute envisioned by the Committee was thus broadened beyond the area of drug research. In line with this development, Herty wrote to A.C.S. President Edgar Fahs Smith in January 1921 to request that the name of the Committee be changed to "Committee on Institute for Chemo-Medical Research" (66). At about this same time, incidentally, P. A. Levene resigned from the Committee for reasons that are not clear, and was replaced by Dr. Carl Alsberg, then Chief of the Bureau of Chemistry, United States Department of Agriculture (66, 67, 68).

By April of 1921, the subcommittee's draft report was ready to be circulated to the larger Committee (69). The major revisions that were made in the draft seem to have resulted from Abel's concern that the report contained too much of a one-sided chemical viewpoint, with the medical side often revealing only a superficial acquaintance with the subject. He feared that there were too many positive statements with respect to therapeutics and fundamental problems in the life sciences that medical scientists could object to (70, 71). Abel's criticisms were taken into account, and the revised report was unanimously approved by the full Committee by early August of 1921 (72). The report was published by the Chemical Foundation in late 1921 as "The Future Independence and Progress of American Medicine in the Age of Chemistry" (73). Herty later commented: "I put more of my heart into that document than into any piece of work in my life" (74).

The report discusses the importance of chemistry in medicine, the need for cooperative research in this area, the lack of adequate facilities in the United States, the superior situation in other countries, and so on. No specific solution is proposed, however, just a plea for cooperative research (73). The efforts of the Chemical Warfare Service, with which several members of the Committee had been associated, were pointed to as an example of productive cooperative research between physical and biological scientists (as discussed by Daniel Jones in his paper in this volume). What happened to the proposed research institute? At first the discussion of the institute and the endowment needed to fund it was pushed back into an appendix, and then it was deleted entirely from the published report.

The reason for this deletion was that Garvan felt that the report would carry more weight if it was directed to the general subject of cooperation between chemists, pharmacologists and biologists, rather than towards the founding of any definite institution. It was decided that specific information on the proposed institute would be printed separately as a special pamphlet and put into the hands of those who indicated an active interest in the subject. This separate pamphlet was to be used as a part of the campaign to secure funding for a chemo-medical institute. The plan that it presented for a research institute was almost identical to that developed by the Committee in 1919 (75, 76, 77).

Almost a million copies of the published report were distributed by the Chemical Foundation to doctors, heads of women's organizations, legislators and other public officials, heads of commercial organizations, educators, libraries, etc. The Foundation spent more than $80,000 in the publication and distribution of the report (78). The object, of course, was to stimulate widespread interest in the need for chemo-medical research.

The Committee continued to function after the publication of the report, but held no meetings nor took any significant action for the next few years. The published report appears to have stimulated interest and commendation, but no funds were forthcoming to establish a research institute (78, 79). In the meantime, Herty had given up his position with the American Chemical Society in 1921 to become President of the newly-created Synthetic Organic Chemical Manufacturers Association. In this position, he could continue to work to promote the development of the synthetic organic chemical industry in this country (12, 13).

Apparently the Committee's report proved useful to a number of universities in fund-raising efforts for chemical research (78, 79). Georgetown University actually did undertake to establish an institute for chemo-medical research on its campus beginning about 1925, and in 1931 initiated such a project, apparently on a rather small scale, with the aid of support from the Chemical Foundation and of Garvan personally (80, 81). The lack of progress in establishing an independent institute of the size and scope envisioned by the Committee, however, must have been somewhat frustrating to Herty.

Then in 1926 an event took place that changed the entire thrust of the Committee's activities. Senator Joseph Ransdell of Louisiana became interested in the proposal for an institute for chemo-medical research and requested a copy of the Committee's report (82, 83). Ransdell had served as Chairman of the Senate Committee on Public Health and National Quarantine, and had been responsible for the establishment of a national leprosarium in his state. He was thus no stranger to public health issues (84).

Ransdell was enthusiastic about the idea of introducing a

bill to establish a government-sponsored medical research institute. Herty told him frankly that the Committee had considered this possibility, but had concluded that a privately-endowed institute would be preferable. He did agree, however, to consult the Committee members about the matter. Ransdell pointed out that there was probably little likelihood of getting such a bill passed in light of the budget situation, but suggested that the publicity generated by the hearings would be useful to the cause (85, 86).

Ransdell was not the first public official to suggest that the Committee turn to the government for support of the proposed institute, but Herty and his colleagues had consistently resisted such suggestions. In 1922, for example, Senator Morris Sheppard of Texas wrote to Herty offering to introduce a bill into the Senate for the establishment of a chemical research laboratory, but the Committee apparently never followed up on this offer (87). When physicist Harvey Curtis of the National Bureau of Standards wrote to Herty in that same year suggesting government sponsorship for the institute, Herty replied that he felt that an endowed institute would be a "better plan" (88). In its published report, the Committee had expressed its opinion that it was unlikely that chemists in government laboratories would be "left free from routine or law-enforcing problems sufficiently long to accomplish much fundamental work." Moreover, the report continued, public funds "are only appropriated for specific purposes and the necessary freedom for work on fundamental principles would, therefore, be practically impossible in a Government institute" (89). That not all of the members of the Committee were so negative about research in government laboratories, however, is suggested by the support readily given to Ransdell's plan by some of them, as noted below.

The Committee's reaction to Ransdell's proposal was mixed. For example, Taylor and Johnson both were very much inclined against any Congressional action, preferring to stay with the plan for a privately-endowed institution. Abel agreed that he would prefer to see the institute established under a private grant, but felt that the publicity that would be generated if a bill were introduced into Congress could help the movement. Stieglitz felt that Ransdell should be encouraged, believing that any research institute with adequate funding was better than none. Hunt and Alsberg, both of whom had worked in government laboratories, strongly favored the Senator's plan (90-95).

Herty communicated the mixed opinions of the Committee to Ransdell. The Senator indicated that he agreed that a privately-endowed institute was desirable, and that his intention was to call attention to the need for such an institution with the hope that some wealthy individuals would step forward to fund it (96, 97). On July 1, 1926, Ransdell did introduce a bill into the Senate calling for the establishment of a National Institute of Health within the United States Public Health Service (98, 99).

Whatever may have been the original intentions of Herty and

Ransdell, the bill to establish a National Institute of Health quickly became for them much more than just an effort to publicize the campaign for a privately-supported research institute. Perhaps Herty was tired of waiting for a rich millionaire to endow his pet project, and finally decided to turn to his Uncle Sam. He may also have been influenced by Francis Garvan. In 1926 Herty left his job as President of the Synthetic Organic Chemical Manufacturers Association to become full-time advisor to the Chemical Foundation. One of the lines of work that Garvan asked him to concentrate on was the "vigorous development of a concerted and thoroughly comprehensive backing for Senator Ransdell's bill" (100). It may be, of course, that Herty interested Garvan in the bill, but having the support of his employer no doubt was an encouragement to him in his efforts on behalf of the bill. In a speech delivered in 1927, Herty now argued that it would not be in the "true American spirit" to await the largess of the rich to fund such an institute. All citizens are subject to illness, he continued, and the financing of medical research is thus a matter for "united action by all our people" (i.e., for tax support) (101).

Over the next several years, Herty campaigned hard for the establishment of the National Institute of Health. He made speeches on the subject, lobbied public officials, and persuaded the American Chemical Society and other organizations to support the bill (101-104). Herty was also one of the key speakers at the hearings that were finally held on the Ransdell bill in 1928, and other members of his Committee also participated in these hearings (105). At the hearings, Herty admitted that he had originally felt that such an institution should not be under government auspices, but that he had since changed his mind. When asked whether his change of opinion was related to the difficulty of obtaining the money for a private institution, he replied: "No; not at all..." (106). One cannot help but wonder, however, whether the Committee's failure to find a private citizen willing to back the venture did not play some role in Herty's change of heart.

The legislative effort was eventually successful, in no small part due to the influence of Herty, and the National Institute of Health established in 1930 (107). Although it was not the privately-endowed research institute that Herty's Committee had in mind, the Committee was apparently satisfied that it could accomplish many of the same purposes. Herty submitted his final report as Chairman to the A.C.S. President in March 1931, and the Commitee was disbanded (108). Although the original goal of establishing an institute of drug research was never accomplished, the Committee's efforts had served to publicize the importance of research in areas such as medicinal chemistry and pharmacology and aided significantly in the campaign to establish the National Institute of Health.

Acknowledgment

This publication was supported in part by NIH Grant LM 03300 from the National Library of Medicine. I wish to thank the following for permission to use archival materials: Special Collections, Robert W. Woodruff Library, Emory University; Alan Mason Chesney Archives, Johns Hopkins University; and Division of Archives, State Historical Society of Wisconsin. Part of the research for this paper was carried out while the author was a Visiting Associate Professor at The Johns Hopkins Institute of the History of Medicine. A preliminary, abbreviated version of the paper was delivered at the American Chemical Society meeting in Washington, D.C., on September 12, 1979, at a session of papers in honor of Aaron J. Ihde sponsored by the Division of History of Chemistry.

Literature Cited

1. [Herty, Charles H.]. J. Ind. Eng. Chem., 1918, 10, 673-674. The quotation is from p. 673.
2. Herty, C. H. to Bailey, J. R. January 23, 1922, Charles Holmes Herty Papers, Special Collections, Robert W. Woodruff Library, Emory University, Atlanta, Georgia, box 74, folder 1. Hereafter this collection will be cited as "Herty Papers."
3. Burdick, Alfred S. J. Amer. Pharm. Assoc., 1922, 11, 98.
4. Haynes, Williams. "This Chemical Age"; Alfred A. Knopf: New York, 1942; p. 75.
5. Beer, John J. "The Emergence of the German Dye Industry"; University of Illinois Press: Urbana, Ill., 1959.
6. Ihde, Aaron J. "The Development of Modern Chemistry"; Harper and Row: New York, 1964; pp. 454-463, 695-699.
7. Arny, H. V. Ind. Eng. Chem., 1926, 18, 949-952.
8. Haynes, Williams. "American Chemical Industry," vol. 3; D. Van Nostrand: New York, 1945; pp. 209-224, 258-259, 311-326.
9. Ibid., pp. 257-278.
10. Drug Chem. Markets, 1920, 6, 686-687.
11. Bud, R. F.; Carroll, P. T.; Sturchio, J. L.; Thackray, A. W. "Chemistry in America, 1876-1976: An Historical Application of Science Indicators" (A Report to the National Science Foundation); Department of History and Sociology of Science, University of Pennsylvania: Philadelphia, 1978; pp. 115-117.
12. Wall, Florence. The Chemist, February, 1932, 123-131.
13. Cameron, Frank. J. Amer. Chem. Soc., 1939, 61, 1619-1624.
14. Wilcox, David, Jr. "American Chemists and Chemical Engineers"; Miles, Wyndham, Ed.; American Chemical Society: Washington, D.C., 1976; pp. 217-218.
15. Herty, C. H. J. Amer. Chem. Soc., 1915, 37, 2231-2246.

16. Browne, C. A.; Weeks, M. E. "A History of the American Chemical Society"; American Chemical Society: Washington, D.C., 1952; pp. 374-376. See also ref. 12, 13, 14.
17. Herty to Abel, John. August 9, 1918, Herty Papers, box 102, folder 1.
18. Herty to Abel, John. August 21, 1918, Herty Papers, box 102, folder 1.
19. Stewart, F. E. J. Amer. Pharm. Assoc., 1919, 8, 256-260.
20. Herty to Abel, John. October 23, 1918, Herty Papers, box 102, folder 1.
21. J. Ind. Eng. Chem., 1918, 10, 969-976.
22. Ibid., 1919, 11, 59-69, 157-161, 377-378.
23. Parsons, Charles to members of Committee on the Institute for Drug Research. February 12, 1919, John J. Abel Papers, Alan Mason Chesney Archives, The Johns Hopkins University, Baltimore, Maryland. Hereafter this collection will be cited as "Abel Papers."
24. J. Ind. Eng. Chem., 1919, 11, 592-593.
25. Herty to Abel. February 19, 1919, Abel Papers.
26. Herty to members of Committee on the Institute for Drug Research. March 11, 1919, Abel Papers.
27. Draft report of Committee on the Institute for Drug Research to President William Nichols. March 15, 1919, Abel Papers. A copy of this report may also be found in the Herty Papers, box 102, folder 3.
28. Abel, John to Herty. November 21, 1918, Herty Papers, box 102, folder 1.
29. Herty to members of Committee on the Institute for Drug Research. March 28, 1919, Abel Papers.
30. Eldred, Frank to Herty. March 5, 1919, Herty Papers, box 102, folder 3. This letter discusses some of the differences in views among Committee members, as expressed at the February meeting.
31. Taylor, Frank to Power, F. B. March 7, 1919, Herty Papers, box 102, folder 3.
32. Haynes, ref. 8, pp. 262-265.
33. Ibid., pp. 265-272.
34. Herty to members of Committee on the Institute for Drug Research. March 4, 1920, Abel Papers.
35. [Herty, C. H.] J. Ind. Eng. Chem., 1919, 11, 718-719.
36. Herty to Francis, J. M. May 2, 1919, Herty Papers, box 102, folder 3.
37. [Leech, P. N.] Chicago Chem. Bull., 1919, 6, 67-68.
38. Leech, P. N.; Rabak, William; Clark, A. H. J. Amer. Med. Assoc., 1919, 73, 754-759. The quotation is from p. 758.
39. Cook, Ellsworth. "The American Society for Pharmacology and Experimental Therapeutics, Incorporated: The First Sixty Years, 1908-1969"; Chen, K. K., Ed.; American Society for Pharmacology and Experimental Therapeutics: n.p., 1969; pp. 177-179.

40. Francis, J. M. to Herty. April 23, 1919, Herty Papers, box 102, folder 3.
41. J. Amer. Med. Assoc., 1915, 65, 67-70. The quotation is from p. 69.
42. Stieglitz, Julius to Herty. May 22, 1919, Herty Papers, box 102, folder 3.
43. Parsons, Charles to Herty. April 17, 1919, Herty Papers, box 102, folder 3.
44. Kremers, Edward to Eldred, Frank. December 11, 1918, Herty Papers, box 102, folder 1.
45. Eldred, Frank to Herty. January 15, 1919, Herty Papers, box 102, folder 2.
46. [Arny, H. V.] J. Amer. Pharm. Assoc., 1918, 7, 1027-1028.
47. Arny, H. V. Ibid., 1919, 8, 455-458.
48. Beringer, George. Ibid., 1914, 3, 1229-1230.
49. J. Amer. Pharm. Assoc., 1917, 6, 1099-2000.
50. Ibid., 1918, 7, 190-192.
51. Ibid., 1920, 9, 335-336.
52. Scoville, W. L. Ibid., 1919, 8, 268.
53. Mrtek, Robert. Amer. J. Pharm. Ed., 1976, 40, 342.
54. Beal, George. J. Amer. Pharm. Assoc., 1919, 8, 260-267.
55. Stewart, F. E. "Prospectus of the American Pharmacologic Society"; American Pharmacologic Society: New York, 1906. Copy in F. E. Stewart Papers, Division of Archives, State Historical Society of Wisconsin, Madison. Hereafter this collection will be referred to as "Stewart Papers."
56. Drugg. Circ., 1905, 49, 259-262.
57. Stewart, F. E. J. Amer. Pharm. Assoc., 1919, 8, 256-260.
58. Stewart, F. E. Proc. Amer. Assoc. Pharm. Chem., 1919, 12, 29-36.
59. Stewart, F. E. Pharm. Era, 1912, 45, 665-667.
60. Stewart, F. E. "Proposed National Institute of Drug Research"; Five-page typescript commentary prepared for Milton Campbell, 1918, Stewart Papers.
61. Haynes, ref. 8, pp. 260-262.
62. Stieglitz, Julius, Ed. "Chemistry in Medicine"; Chemical Foundation: New York, 1928, p. IX.
63. Herty, C. H. to members of Committee on the Institute for Drug Research. September 30, 1920, Abel Papers.
64. "Minutes of the Joint Meeting [Oct. 11, 1920] of the Committee on Institute for Drug Research and Representatives of the Chemical Foundation, Inc."; Three-page typescript, Abel Papers.
65. Herty, C. H. to members of Committee on the Institute for Drug Research. November 13, 1920, Abel Papers.
66. Herty, C. H. to Smith, E. F. January 26, 1921, Herty Papers, box 102, folder 5.
67. Levene, P. A. to Stieglitz, Julius. December 14, 1920, Herty Papers, box 102, folder 5.

68. Stieglitz, Julius to Herty. November 29, 1920, Herty Papers, box 102, folder 4.
69. Herty to members of Committee on Institute for Chemo-Medical Research. April 9, 1921, Abel Papers.
70. Abel, John to Herty. May 23, 1921, Abel Papers.
71. Abel, John to Hunt, Reid. May 23, 1921, Abel Papers.
72. Herty to members of Committee on Institute for Chemo-Medical Research. June 20 and August 5, 1921, Abel Papers.
73. Abel, John; Alsberg, Carl; Bacon, Raymond; Eldred, F. R.; Hunt, Reid; Johnson, Treat; Stieglitz, Julius; Taylor, F. O.; Herty, Charles. "The Future Independence and Progress of American Medicine in the Age of Chemistry"; Chemical Foundation: n.p., n.d.
74. Herty to Lawrence, William. May 2, 1924, Herty Papers, box 102, folder 8.
75. Draft of report of Committee on Institute for Chemo-Medical Research, undated, Abel Papers. This draft includes the plan for the institute as an appendix.
76. Herty to members of Committee on Institute for Chemo-Medical Research. January 27, 1922, Abel Papers.
77. "Appendix," Herty Papers, box 102, folder 10. This is a two-page printed document describing the proposed institute, the pages being numbered 85 and 86. Apparently this was originally intended to be bound in with the published report of the Committee (ref. 73), which ends with page 83. In the actual published version, however, blank pages labelled "Comments" are numbered as pages 84-96 and the appendix does not appear.
78. Mead, Larkin to Herty. March 16, 1923, Herty Papers, box 102, folder 8.
79. Annual Reports of the Committee on Institute for Chemo-Medical Research to President of A.C.S., 1923-1925, Abel Papers.
80. Lyons, C. W. to Herty. January 31, 1925, Herty Papers, box 102, folder 9.
81. "Notebook; Chemical Foundation, Inc."; vol. 2, pp. 67, 74. Francis P. Garvan Estate Collection, American Heritage Center, University of Wyoming, Laramie, Small File 13. I am indebted to P. Thomas Carroll for calling my attention to this reference.
82. Herty to Ransdell, Joseph. January 19, 1926, Herty Papers, box 103, folder 3.
83. Ransdell, Joseph to Herty. February 5, 1926, Herty Papers, box 103, folder 3.
84. Furman, Bess. "A Profile of the United States Public Health Service, 1798-1948"; United States Department of Health, Education and Welfare: Washington, D.C., 1973; pp. 308-311, 374.
85. Ransdell, Joseph to Herty. February 27, 1926. Herty Papers, box 103, folder 3.

86. Herty to members of Committee on Institute for Chemo-Medical Research. March 22, 1926, Abel Papers.
87. Sheppard, Morris to Herty. February 18, 1922, Herty Papers, box 102, folder 7.
88. Herty to Curtis, Harvey. December 12, 1922, Herty Papers, box 102, folder 7.
89. Abel, et al., ref. 73, p. 67.
90. Taylor, F. O. to Herty. March 26, 1926, Herty Papers, box 102, folder 9.
91. Johnson, Treat to Herty. March 26, 1926, Herty Papers, box 102, folder 9.
92. Abel, J. J. to Herty. March 23, 1926, Abel Papers.
93. Stieglitz, Julius to Herty. March 26, 1926, Herty Papers, box 102, folder 9.
94. Hunt, Reid to Herty. March 25, 1926, Herty Papers, box 102, folder 9.
95. Alsberg, Carl to Herty. March 30, 1926, Herty Papers, box 102, folder 9.
96. Herty to Ransdell, Joseph. April 12, 1926, and April 15, 1926, Herty Papers, box 103, folder 3.
97. Ransdell, Joseph to Herty. April 5, 1926, Herty Papers, box 103, folder 3.
98. Ransdell, Joseph to Herty. July 1, 1926, Herty Papers, box 103, folder 3.
99. "The Ransdell Bill"; memorandum from Herty to Francis Garvan, May 14, 1930, Herty Papers, box 104, folder 4.
100. Herty to Garvan, Francis. January 26, 1927, Herty Papers, box 103, folder 4.
101. Herty, C. H. "The Ultimate Mission of Chemistry: Good Health" (address delivered in Richmond, Virginia, on October 21, 1927); Chemical Foundation: New York, n.d.
102. Letters from Herty to Committee on Institute for Chemo-Medical Research. March 12, 1927, and June 5, 1928, Abel Papers.
103. Herty to Cumming, Hugh. May 26, 1927, Herty Papers, box 103, folder 5.
104. Herty to Pierce, C. C. May 3, 1927, Herty Papers, box 103, folder 5.
105. Report of Senate Committee on Commerce on S. 4518, May 3, 1928; pp. 8-22, 28-42, 71-76, 92-96.
106. Ibid., p. 72.
107. Harden, Victoria Angela. "Charles Holmes Herty, Joseph Eugene Ransdell and the Establishment of the National Institute of Health, 1926-1930"; unpublished manuscript. This paper provides a detailed discussion of the campaign for the N.I.H. bill. I am indebted to the author for a copy of her paper.
108. Herty to members of committee on Institute for Chemo-Medical Research. March 9, 1931, Abel Papers.

RECEIVED December 16, 1982

6

Sulfanilamide and Diethylene Glycol

JAMES HARVEY YOUNG
Emory University, Department of History, Atlanta, GA 30322

In 1937 the S. E. Massengill Company in Tennessee marketed a liquid dosage form of sulfanilamide, unaware that the solvent used, diethylene glycol, was poisonous. Before the Food and Drug Administration could track down this "Elixir Sulfanilamide", 107 persons had died. The manufacturer was assessed the largest fine ever levied under the Food and Drugs Act of 1906. The crisis prompted research in the FDA's division of pharmacology which developed a statistically based method for determining the comparative toxicity of compounds. The episode was also responsible for a provision in the Food, Drug, and Cosmetic Act of 1938 requiring that the safety of new drugs must be established prior to marketing, a precedent later applied to pesticides, food and color additives, and medical devices.

In a promotional brochure sent to physicians in October 1937, the S. E. Massengill Company of Bristol, Tennessee, announced "A New Sulfanilamide" (1). "Our research department has just released an Elixir Sulfanilamide," the report read, "40 grs to the fluidounce. It is ideal for your patients who can take liquids--but little else. Also, it is not unpleasant to take, so is suitable for children. The color is brilliant red. . . ."

Thus began the marketing of a therapeutic solution that mixed two chemical compounds, one the first drug to combat effectively bacterial infections, the other a powerful poison. The prescribing of Elixir Sulfanilamide caused over a hundred deaths, many of them of children; provoked national panic; and spurred a desperate effort, led by the Food and Drug Administration, to track down the lethal "elixir" and to remove it from drugstores and homes. In the longer run, the Elixir Sulfanilamide disaster left its mark on regulatory law, and hence on pharmaceutical manufacturing, and prompted significant innovation in toxicology.

0097–6156/83/00228–0105 $06.25/0

Sulfanilamide

By the mid-1930s most researchers and clinicians concerned with infectious diseases had abandoned hope that chemical agents would ever be discovered to provide effective treatment for bacterial infections (2, 3). The optimism of the teens, spurred by Paul Ehrlich's pioneering chemotherapy for syphilis, had led to many efforts in Europe and America to develop chemical synthetic drugs that would combat germs in the body, but this optimism had waned, battered by a series of failures, drugs glowingly announced--especially in Germany--that with experience had not proved out. Emil von Behring's old dictum had regained currency: "inner disinfection is a vain dream" (2).

Therefore, Gerhardt Domagk's announcement of Prontosil in 1935 (4, 5) sparked scant initial interest and the usual skepticism. The British bacteriologist Ronald Hare assumed Prontosil to be "[a]nother of those damned compounds from Germany with a trade name and of unknown composition that are no use anyway" (6). Domagk's data were too sketchy, his perfect results in animal experiments too pat, the accompanying clinical reports too resonant with a testimonial tone (2).

French scientists at the Pasteur Institute, however, promptly dispelled some of Prontosil's mystery, splitting the molecule into a red dye component and an old chemical, sulfanilamide, its 1909 patent long since lapsed (2). The suspicion arose that Domagk, an I. G. Farbenindustrie researcher, had rediscovered sulfanilamide, and that the manufacturer had held it off the market until it could be presented in a new, complex, disguised, and patentable form (2, 3, 6). Whether or not the suspicion was true, the French scientists, by showing that sulfanilamide was the therapeutically active fraction of Prontosil, shattered the gigantic Germany company's profitable plans.

Persuasive proof of sulfanilamide's true antibacterial prowess came from Leonard Colebrook and his team of researchers in the maternity ward of Queen Charlotte's Hospital in London (7, 8, 9). Managing to obtain samples of Prontosil late in 1935, Colebrook quickly demonstrated the drug's remarkable effectiveness in curbing puerperal fever. The utility was soon expanded to other severe streptococcal infections.

Colebrook announced his promising results during the summer of 1936 at an international congress of microbiology in London (3, 6). Perrin H. Long of the Johns Hopkins University, attending the conference, learned of Colebrook's presentation and cabled the news to his colleagues. On September 1 he and Eleanor A. Bliss began experiments with mice, with such promising results that a week later Long treated a child desperately ill with erysipelas, a woman with infection following an abortion, and then many more severe cases of streptococcal infection (3, 10, 11). Long and Bliss had not yet published their encouraging results when Long received a telephone call from Eleanor Roosevelt (3).

Franklin D. Roosevelt, Jr., his mother said, lay desperately ill in Massachusetts General Hospital. The doctors thought he might be dying. An infection in his throat had spread to his sinuses and was now, carried through his blood, engulfing his body.

A week later, on December 16, 1936, young Roosevelt's physician, George Loring Tobey, Jr., told reporters that his patient had responded "beautifully" to treatment with a trade form of sulfanilamide and was now out of danger (12). His mother and fiancee Ethel duPont had left his bedside for the first time in nearly three weeks. Headlines the nation over carried the news. "Young Roosevelt Saved by New Drug," reported the *New York Times*, and the next day, in an editorial, heralded "DOMAGK's discovery as the outstanding therapeutic achievement of the last decade" (13). Under similar circumstances a decade earlier Calvin Coolidge, Jr., had died of streptococcal infection (3). Franklin D. Roosevelt, Jr., survived, recovered, graduated from Harvard the next June, and a week later married Miss duPont (14, 15).

The Roosevelt case helped trigger the "explosive suddenness" (16) with which American physicians and the American public generally developed a widespread interest in sulfanilamide. Chemical manufacturers began to produce the raw drug and sell it to pharmaceutical firms which processed it into pills and capsules for distribution to physicians and pharmacists (17). At least a hundred firms were making proprietary brands of sulfanilamide tablets in 1937, among them the Massengill Company. Use of the drug quickly assumed tremendous proportions. Public interest quickened further with the report that sulfanilamide could cure gonorrhea, news that surfaced amidst a major Public Health Service campaign against venereal disease (18). This led, despite PHS warnings, to widespread self-dosage with sulfanilamide secured from druggists without a physician's prescription. The hazards of such self-dosage became increasingly apparent as adverse reactions which could result from sulfanilamide's use began to be recognized and reported in the medical literature (19).

The majority of streptococcal infections occurred in children under ten, so the new drug proved a particular blessing to the very young (20). The big tablets of sulfanilamide could be administered successfully in hospitals to all but the smallest babies. Most sick children, however, received treatment at home, and mothers found it difficult to get them to swallow large pills. This circumstance seemed to call for a liquid dosage form. A number of attempts to find a suitable vehicle, however, proved unavailing.

Diethylene Glycol

Success at finding a solvent for sulfanilamide, which had eluded others, quickly crowned the efforts of Harold Cole Watkins, chief chemist, chief pharmacist, trouble-shooter, and developer of new formulas at the Massengill Company (21, 22). Learning from a company salesman that a good market awaited a liquid dosage form, Watkins set to work in July 1937 and in a few days developed a formula. He knew from his reading that glycols possessed remarkable solvent properties, and the company had found this class of chemicals satisfactory, used in very small amounts, in several other drugs. So Watkins sought to dissolve sulfanilamide in diethylene glycol and found that he could dissolve as much as 75 grains in a single fluidounce. When chilled, however, the drug tended to separate out. So Watkins decided on a ratio of 40 grains per ounce. Mixing 80 gallons of water and 60 gallons of diethylene glycol, Watkins added 58 1/2 pounds of sulfanilamide powder, plus small quantities of elixir flavor, saccharin, caramel, amaranth solution, and raspberry extract. He then sent his "Elixir Sulfanilamide"--akin in color to the original Prontosil--to the company's "control laboratory," where the new product was checked for appearance, flavor, and fragrance, and approved. Commercial distribution began from Bristol on September 4, and shortly afterwards from the company's branch plant in Kansas City, to which the formula had been sent (23).

Watkins had made no tests whatsoever to assess the toxicity of either the separate ingredients or the combination of ingredients in the final formula (21, 22). He later stated that there was no point in making tests, since the action of the ingredients was well understood (22). The glycols, related to glycerine, had been widely used by drug companies, and, Watkins averred, were well known not to be toxic.

Theodore Klumpp, the Food and Drug Administration's chief medical officer, who interviewed Watkins soon after disaster struck, expressed amazement at the chemist's "revelation of ignorance" (22). For anyone alert to the literature, warning signs abounded.

Ethylene glycol was first prepared in France in 1859, and diethylene glycol soon thereafter, but prior to 1925, when a plant was erected in West Virginia to produce glycols, these chemicals possessed little commercial importance (24). Their main uses were to lower the freezing point of dynamite, to cool airplane engines, and to protect automobile radiators from freezing. Diethylene glycol, however, did not work well as an antifreeze: its success as a solvent, which Watkins prized, made it unsuitable, for if spilled it would dissolve a car's paint. The hygroscopic properties of diethylene glycol gave it a role as humectant for tobacco, printing ink, glue, cellophane, and, in minute quantities, for toothpaste and drug ampules.

For some years prior to Watkins' employment of diethylene glycol, the Food and Drug Administration had advised against the use of glycol solvents in foods, declaring that definite, comprehensive conclusions as to the physiological action of these chemicals could not be reached on the basis of existing scanty research (25). The primary manufacturer, Carbon and Carbide Chemicals Corporation, had consistently discouraged use of diethylene glycol in either food or drugs (26). Beginning in 1931 more explicit reports of the poisonous nature of the chemical appeared in medical journals (27, 28). The point about Watkins, however, was not just that he did not know the relevant literature; he did not take elementary precautions and test for toxicity.

Dawn of Disaster

On October 11, American Medical Association officials in Chicago received two telegrams, one from the president of the Tulsa County Medical Society in Oklahoma, the other from the Springer Clinic of Tulsa (23, 29, 30). Six deaths had occurred following the administration of Elixir Sulfanilamide-Massengill. What, the telegrams inquired, was the composition of this drug?

The AMA replied immediately. No product made by the Massengill Company had been accepted by the AMA's Council on Pharmacy and Chemistry, a body created in 1905 to test the claims of proprietary medicines sold or prescribed through physicians (23, 31). Nor had the Council recognized any solution of sulfanilamide. Simultaneously, the AMA telegraphed the Massengill Company in Bristol requesting the formula (23). The company complied, asking, however, that the composition be regarded as a trade secret and be kept confidential. Diethylene glycol became immediately suspect, its presence verified by the AMA in the first sample of the death-dealing drug rushed from Tulsa to Chicago. The AMA also secured a gallon of the Elixir from Massengill and bought more on the open market in Tulsa, which it used in preliminary animal experiments to demonstrate the drug's poisonous potential.

Deaths in Tulsa mounted, and a St. Louis pathologist reported a series of deaths across the Mississippi River in East St. Louis (32). Morris Fishbein, editor of the Journal of the American Medical Association, wrote a warning to appear in the October 23 issue (29), and on October 18 met with reporters to give them an advance copy of the text (23). The press and radio quickly alerted the nation to the catastrophe in progress. The AMA arranged for more probing research regarding Elixir Sulfanilamide's toxicity, turning to Perrin Long at Johns Hopkins and his colleague, E. K. Marshall, Jr., who had discovered sulfanilamide's mode of action in the body, and to E. M. K. Geiling, Paul R. Cannon, and E. M. Humphreys of the University of

Chicago (2, 23). The AMA also established liaison with the Food and Drug Administration.

FDA had received its first word of the Tulsa deaths on October 14 in a telephone inquiry from a New York physician associated with a pharmaceutical company (23). Immediately the agency sent inspectors to Tulsa from Kansas City (33). They learned that suspicions had surfaced on October 7 when autopsy cases began to accumulate, all revealing kidney degeneration, with anuria having preceded death. The symptoms resembled mercuric chloride poisoning, except for the absence of enteritis and liver damage. Elixir Sulfanilamide had been found to be the drug common to the nine deaths so far discovered, eight of them among children being treated for strep throats, the ninth an adult male with gonorrhea. The county medical society had appointed a committee of inquiry and had urged the Tulsa Retail Druggists Association to suspend selling Elixir Sulfanilamide. A Tulsa pathologist had confirmed the drug's toxicity on guinea pigs.

The inspectors had been invited to attend a postmortem examination of the adult man. His kidneys were "excessively enlarged," about fifty percent larger than normal. "Clotted, purplish congestion" appeared "on [the] periphery of [the] kidneys." Samples of Elixir Sulfanilamide collected by the inspectors were sent to Central District headquarters in Chicago (33).

The S. E. Massengill Company

The first reports from Tulsa led the FDA to dispatch the agency's chief medical officer, Theodore G. Klumpp, accompanied by an inspector, William T. Ford, from Cincinnati, to the plant in Bristol at which Elixir Sulfanilamide had been conceived and first prepared (21, 22).

The S. E. Massengill Company, the largest pharmaceutical company in the South, had begun in 1897 as a partnership between two brothers, one of whom had died in 1926 (34, 35). The survivor, Samuel Evans Massengill, a man in his sixties, continued to manage the firm. Massengill admired his ancestry: in 1769 a direct forebear had become the second permanent white settler in what would later be Tennessee, and Massengill's father had served in the Confederate cavalry before attending medical school and living out his life as a country doctor and raiser of gaited horses.

Massengill himself had been graduated from the University of Nashville Medical School, but had not practiced, launching his drug manufacturing firm instead (37). This venture had prospered. By 1937 the firm made a wide array of drugs for human and animal use, including many private formulas for physicians and druggists (35, 36). Manufacturing was done in the Bristol and Kansas City plants, and the company also had distributing offices in New York and San Francisco (34). Some five hundred employees worked for Massengill, including a corps of 160 detail men to make its products known. Massengill drugs went overseas, especially to Latin America and Egypt (35). Prompt service was a Massengill hallmark. S. E. Massengill enjoyed an excellent reputation in his community, and Bristol civic leaders and physicians staunchly supported him in his time of troubles which had just begun (38).

Through most of the years that the Food and Drugs Act of 1906 had been in effect, enforcers of that law had paid little heed to the Massengill Company's existence (36). No thorough inspection by investigational procedures had preceded the Elixir Sulfanilamide crisis. The company itself deemed its record under the law "splendid," more admirable than that of most drug manufacturers (34). Only three cases had been brought against the company by FDA, all during the 1930s. Two prosecution cases the company settled by paying small fines, $250 and $150, rather than undertaking the expenses of a trial (39, 40). These involved Fluidextract of Colchicum, overstrength when judged by the official standard of the National Formulary, and Tincture of Aconite, which the FDA found understrength. The third case, concerning a seizure of Elixir Terpin Hydrate and Codeine, went to trial as both adulterated and misbranded, the government finding a disparity between the quantity of both ingredients stated on the label and actually present in the medicine (41, 42). The judged ruled for the government.

When Dr. Klumpp and Inspector Ford met Dr. Massengill in his Bristol office, the proprietor "looked worried" and offered his "full cooperation" (22). He had already done "all that was humanly possible," Massengill said, to recall outstanding stocks of Elixir Sulfanilamide. Some 375 telegrams had been sent from Bristol, and more from other company branches, a total of 1100, to the firm's salesmen, who had distributed small sample bottles, and to concerns, mostly drug wholesalers, whose orders had been filled. Klumpp insisted, since these recall telegrams had given no hint of hazard, that Massengill dispatch follow-up telegrams reading: "Imperative you take up immediately all elixir sulfanilamide you may have dispensed. Product may be dangerous to life. Return all stocks at our expense" (23).

That the diethylene glycol in his Elixir Sulfanilamide had caused the deaths being increasingly reported, Dr. Massengill, in his conversation with the FDA officials--or thereafter--, did not admit. No special toxicity tests had been made on the Elixir, he acknowledged, indeed, no animal tests at all (21, 22). Without specifying details, Massengill asserted that his control department saw to it that all drugs were "all right" before their release. The real blame lay, he suggested, not with the solvent but with the sulfanilamide.

Massengill stated, as Klumpp reported the conversation (22), "that the drug sulfanilamid[e] had been so exploited by physicians and the press that everyone in the country was going wild with it and using it for everything and now the disastrous effects of it were coming out. . . . [T]hey don't know much about sulfanilamid[e] but that it was well recognized that it was dangerous when used with other drugs. He stated his opinion that the fatalities from Elixir Sulfanilamide were due not to the Elixir Sulfanilamide itself, but to the fact that those that were killed took other drugs while taking the Elixir Sulfanilamide, the combination of which killed them." In later public statements Massengill adhered to sinister synergism as his explanation of the disaster. Within a few days he told the press: "I do not feel that there was any responsibility on our part" (23). And he wrote the AMA: "I have violated no law" (23).

Klumpp and Ford also talked with Watkins (21, 22). Besides finding out how he had happened to hit upon the Elixir Sulfanilamide formula, they probed his background as a chemist. He had been graduated from the University of Michigan School of Pharmacy in 1901, Watkins said, with the degree of pharmaceutical chemist, and afterwards had experienced a peripatetic career. He had done research in alkaloids for several years, then worked for a while in the analytical laboratory of Merck, followed by short stints in various wholesale and pharmaceutical houses, many of them in western states. In 1929 he had signed a stipulation with the Post Office Department agreeing to abandon the sale of a weight-reduction remedy which he was promoting with excessive claims (23). During the depression, Watkins had been idle some of the time, before joining Massengill.

What impressed Klumpp and Ford most about Watkins was what they deemed a certain callousness in his conversation. He spoke of a preparation of colloidal sulfur he had devised. When marketed, this compound resulted in the death of a number of people. "Mr. Watkins told about this event," Klump wrote, "as if it were an ordinary incident in the business of making and marketing pharmaceuticals." Watkins was getting ready to release a product containing another drug, cinchophen, about which the FDA had issued strong warnings.

"It would seem," the FDA's head physician concluded (21), generalizing from Watkins' words to the broader pharmaceutical marketplace, "that a great many drugs are being compounded and placed upon the American market without adequate testing. The expressed or implied attitude of certain drug manufacturers seems to be that drugs can be tested on the American public. If they fail to kill, or injure in such a way that the injury can be detected and traced to its source, the products have then met their trials successfully. The conclusion is unescapable that such drug manufacturers are perfectly willing to wait for reports of death or injury for information concerning the toxicity of their drugs."

Watkins told Klumpp and Ford that, when news of the deaths had first reached him, he had himself taken a huge oral dose of diethylene glycol, without ill effects. Klumpp doubted Watkins' story, but even if it were true, Klumpp observed, such a "futile, heroic gesture" after the fact could not counterbalance the failure to test the drug properly prior to its distribution.

The Massive Search

The 1906 law did not require premarket testing. Nor did it provide the Food and Drug Administration with authority to seize drugs because they were dangerous. Indeed, only an odd happenstance gave the FDA legal sanction to track down and remove the lethal liquid from drugstores, physicians' offices, and the homes of patients for whom it had been prescribed (23). The company had named this form of sulfanilamide an "elixir," implying an alcoholic solution. Since it contained diethylene glycol instead of alcohol, it was misbranded. "Had the product been called a 'solution,' rather than an 'elixir,'" Secretary of Agriculture Henry A. Wallace said, "no charge of violating the law could have been brought."

The company's telegrams prompted the return of a large quantity of the "elixir" to the manufacturer, where it was taken under state or federal control (23). To secure the remainder of the 240 gallons which had been produced, the FDA launched the largest and most zealous search that had been undertaken during the three decades that the 1906 law had been in force. Almost the entire field force of 239 inspectors and chemists participated. State authorities, particularly in states having drug inspectors, and officials in a number of cities, provided a certain amount of help, but some FDA agents thought greater assistance might have been rendered (43).

The quick action accounted for most of the "elixir." Toward the end of November, Secretary Wallace reported to the Congress, of the 240 gallons produced, 228 gallons and two pints had been "seized . . ., destroyed, collected as laboratory samples, or wasted by spillage and breakage" (23). (Twelve years later a bottle that had been missed turned up in a batch of old medicines taken from a defunct Texas clinic.) (44) Eleven gallons and six pints of Elixir Sulfanilamide had been dispensed, mostly on the prescription of physicians, a small amount by the over-the-counter sale of druggists (23). About half of the "elixir" reaching patients had been consumed; the other half was retrieved before it could be taken.

Finding that "elixir" in the possession of the sick, or in the hands of survivors of those who had died, led to many adventures on the part of food and drug officials, and gave them a gloomy picture of the state of health care in the nation. At the start, getting records from the Massengill detail men posed some difficulties. Most were cooperative in indicating where sales had been made and physicians' samples left (23). But some salesmen proved to be elusive, and a few reluctant to provide needed information. A detail man in Texas had to be jailed by state authorities before he would reveal his record of sales.

Searchers for the dangerous "elixir" often worked to the borders of exhaustion. A tired Alabama state inspector signed a check "Elixir Sulfanilamide" (45, 46). In Atlanta, an elderly FDA inspector, who also was ill, drove through the rain into the north Georgia mountains to a drugstore that could not be reached by phone (47, 48). A pint bottle of the "elixir" had been sent to the druggist, and the inspector was charged with bringing it back. The druggist had the bottle, but four ounces of its red contents were missing, all prescribed by a doctor for one patient, name unknown. When the physician returned from his rounds, he said that he kept no records but believed he had prescribed the medicine for a woman named Lula Rakes. He did not know where she lived, and her surname was not uncommon in the region. The inspector, shouting an inquiry to the deaf druggist, attracted the attention of a bystander, who volunteered the guess that Lula lived eight miles over the ridge in Happy Hollow. The inspector drove over the dark mountain road, only to find Lula's home abandoned. A neighbor said that the Rakes' family had moved one valley farther on. Driving onward the inspector at last found the house, and Lula was there. Busy with the moving, she had taken only a few doses of her medicine, but she could not recall what she had done with the bottle. An hour of conversation and exploring finally led to the medicine, with many other articles in a paper sack under a bed. The inspector began his weary way home.

Other tales had less happy endings. In South Carolina an inspector found on the grave of a black man a partly used bottle of Elixir Sulfanilamide, bearing the name of the physician who had denied having prescribed it (49). In accordance with Gullah custom, the medicine--as well as dishes, spoon, and a bottle of catsup--which he had been using when he died accompanied him to his final resting place.

In Arkansas an inspector found among drugstore prescriptions that two ounces of the "elixir" had been sold to "Jewell" Long (45). Both the druggist and the physician said they knew no such person, the latter asserting he must have treated an itinerant laborer. After searching out a number of Long families without success, the inspector arrived in the neighborhood of the next Long on his list, to learn that a Long girl, seven years of age, named "Jenell," had died the day before, and her funeral was at that moment under way in her home. The inspector went to the house, waiting an opportunity to talk with the bereaved parents. The physician who had denied knowing "Jewell" drove up to the house.

Other physicians and druggists, concerned about their reputations and about damage suits, turned up playing cowardly or villainous roles in the somber "Chronological file of Inspectors' & others' reports regarding deaths caused by the Elixir," compiled in FDA headquarters (50). One doctor went to a drugstore and extracted and destroyed a prescription he had given for the fatal "elixir" (43, 49). A druggist scratched out from their prescriptions the names of three patients who had died.

Such incidents, however, were more than counterbalanced by examples of earnest endeavor to aid in the search for Elixir Sulfanilamide prescribed in good faith. One physician postponed his wedding to help track down the "elixir" prescribed for a boy of three who had moved with his family into the mountains (45). This quest proved successful, in time to save the lad. Many physicians shared the remorse of a New Orleans doctor: "Nobody but Almighty God and I can know what I have been through in these past few days. . . . [T]o realize that six human beings, all of them my patients, one of them my best friend, are dead because they took medicine that I prescribed for them innocently, . . . as recommended by a great and reputable pharmaceutical firm in Tennessee; well, that realization has given me such days and nights of mental and spiritual agony as I did not believe a human being could undergo and survive" (23).

All those who persisted in taking the prescribed "elixir" beyond the onset of initial symptoms until the dosage led to their deaths suffered "stoppage of urine, severe abdominal pain, nausea, vomiting, stupor"; some went into convulsions (23). One hundred seven deaths were reported among patients, a high proportion of them children, who had taken Elixir Sulfanilamide (51, 52). The death toll spanned the country from the Atlantic to the Pacific, although most of those who died lived in the South (23). Had the entire 240 gallons been taken as medicine, it was estimated, some four thousand victims would have lost their live (53).

The Judgment

Besides seizing Elixir Sulfanilamide wherever it could be traced, the Food and Drug Administration saw to the filing of criminal charges against Dr. Massengill and his firm, both in Knoxville and in Kansas City (54). Because he had conceded two earlier prosecution charges, he was a second offender and thus subject to more severe sanctions under the 1906 law.

Massengill termed the pending legal actions "ridiculous" (34). Because of the many counts and his status as a second offender, Massengill said, he could be fined $261,000 and sent to jail for 261 years. By its "continuous gouging," the FDA seemed determined to secure his "financial ruin."

Considering himself "blameless" in the episode (55), Massengill issued a series of documents offering his defense. In "A Description of Glycols" and "The Facts about Sulfanilamide" Massengill downplayed the toxicity of diethylene glycol and blamed sulfanilamide's interaction with other drugs for deaths that had occurred (56, 57). Final results would show, he insisted, that many more people had been aided by his product than had been killed by it. He claimed credit for the prompt and complete removal of Elixir Sulfanilamide from the market and blamed the press for oversensational reporting in order "to unfairly create prejudice" and to injure the company. No "error had been made in compounding the formula." Indeed, "few, if any," pharmaceutical manufacturing plants in the United States were "better suited" than Massengill for their task.

Nonetheless, while making such protestations of rectitude, the company fired Watkins, acknowledged a certain moral responsibility, and settled out of court a number of suits brought by the bereaved families of victims (57, 58). At one stage in this process, Time reported that the highest amount paid up to that point had been $2000 (37). Later the FDA reported a rumor that half a million dollars had been expended in settling damage suits (55).

When, following normal procedures, the FDA offered the Massengill Company an opportunity for a hearing before the chief of the Cincinnati Station, no one appeared in person (59). Instead, company attorneys sent a written reply to the citation, asserting that no violation of the law had occurred. Later Walter G. Campbell, chief of the Food and Drug Administration, offered the lawyers an opportunity to make an appearance before him and provide additional statements, but it was not accepted.

Campbell and his staff anticipated that the battle in court would be hard fought. They knew that company agents had been active gathering affidavits from patients who had used Elixir Sulfanilamide and who would testify that they had not been harmed (55). Some would say that the medicine had helped them. Perhaps as many as 680 such documents, the FDA believed, had been collected (60). It was also anticipated that the company intended to assert that sulfanilamide afflicted the same proportion of patients who took it alone as had been afflicted while taking it mixed with diethylene glycol, about fifteen percent in both circumstances.

FDA officials, believed, however, that their own case would prove to be more persuasive. Granted its legal basis was somewhat narrow. The government would assert that the drug "fell below the professed standard under which it was sold," in that Elixir Sulfanilamide had been labeled to possess the same therapeutic action as sulfanilamide, whereas "its principal action was that of an acute poison" (61-65). Therefore, the term "Quality Pharmaceuticals," as well as the word "Elixir," on the labels constituted misbranding and adulteration. The magnitude of the disaster broadened the thrust of the government's charge. Nor was there any doubt that diethylene glycol had been the fatal component. The nation's leading scientific experts, first brought into the affair by the AMA, would so testify, on the basis of continuing experiments. These included Professor Long of Johns Hopkins and Professors Geiling and Cannon--as well as Anton

J. Carlson, a frequent and skillful witness in FDA's behalf--of Chicago. FDA's own scientists also had been working overtime preparing for the trial.

In September 1938 Dr. Massengill's attorneys filed a demurrer to the government's charges, claiming that he had not violated the law. The judge overruled this plea (54, 65). When Massengill appeared in court, on the day his trial was scheduled to begin, October 3, he changed his plea from not guilty to guilty on a majority of the counts in the charge against him, saying he would do likewise in the case involving shipments of the "elixir" made from Kansas City. Massengill was fined $150 each on 174 counts, the $26,100 total being the largest fine ever assessed under the 1906 law. FDA officials, who did not press for Massengill's imprisonment, expressed themselves as being "much gratified at the outcome" (66).

When William Ford last talked with Harold Watkins, idle at home, the chemist asked the FDA inspector to keep him in mind if Ford "heard of anyone wanting a man" (58). Before the trial, Watkins took his own life (67).

Impact upon Scientific Research

The Elixir Sulfanilamide episode had a marked influence upon research in both industry and government. In the 1930s, the vanguard among American pharmaceutical manufacturers had reached a new plateau of scientific competence. Earlier, while considerable research took place in company laboratories, much of it was held in low esteem by physicians and academics (68, 69). Scientists who moved from universities to industry were expelled from membership in the American Society for Pharmacology and Experimental Therapeutics (70). The new wave of enhanced scientific achievement in industry may be exemplified by the plans initiated by George W. Merck. In 1933 Merck established a Research Laboratory at Rahway, New Jersey, possessing a campus atmosphere and offering interdisciplinary teams of scientists "the greatest possible latitude and scope in pursuing their investigations, the utmost freedom to follow leads promising scientific results no matter how unrelated to what one would call practical returns," as well as freedom to publish (71, 72). Other major companies followed suit, vastly augmenting manpower and equipment devoted to drug research.

The arrival of sulfanilamide accelerated these trends (73). Pharmaceutical chemists employed by manufacturers plunged into research on derivatives and associated preparations, quickly discovering more effective analogues. The old dream of using chemistry to devise synthetic drugs capable of combatting disease germs in the body had been converted into reality. Sulfanilamide helped provide "proof that the impossible had become feasible" (74), stimulating a much broader quest for chemotherapeutic agents. Elixir Sulfanilamide underlined the need to consider safety as a primary factor in the search. Soon the law would make safety a sine qua non.

The Elixir Sulfanilamide crisis placed urgent research pressures on Food and Drug Administration scientists. Glycol solvents played a considerable role in food flavorings, for example, so more research as well as strenuous recall efforts followed news of the deaths from diethylene glycol (25, 75, 76). These efforts expanded into a closer scrutiny given to all types of solvents, diluents, and excipients used in making drugs.

The major FDA concern came to be better comprehension of diethylene glycol's toxicology. The imminent trial in court required this. In a more basic sense, the crisis made FDA scientists aware of inadequacies in the state of the discipline. In constant contact with their peers at the AMA and at the University of Chicago and Johns Hopkins, a team of FDA scientists launched a project that "developed the first valid process for determining the comparative toxicity of compounds, a statistically based and legally defensible process that opened the door to modern toxicological testing methods" (77).

A Division of Pharmacology had been formally set up in the Food and Drug Administration in 1935, composed mostly--as one of its members, Edwin P. Laug, remembered--of "biochemists who then changed sails and became pharmacologists" (78). To study the toxicity of lead and arsenic pesticide residues formed the division's initial purpose, but the Elixir Sulfanilamide crisis brought an almost total shift of effort to diethylene glycol.

The concept of Lethal Dose 50 had already been coined, developing from innovative research in England by J. W. Trevan (79) and in America by Torald Sollmann (78). Yet that concept was as yet rudimentary. "[I]n those days," as Dr. Laug put it without too much exaggeration, "if you had two rats and you killed one and didn't kill the other one, then that was called an LD 50." The Elixir Sulfanilamide disaster made these FDA scientists realize that a statistically significant way of defining LD 50 was needed, so that two poisons could be accurately compared. Herbert E. Calvery gave great impetus to this quest and brought in the statistician Chester I. Bliss (80) as a part-time consultant. Bliss was an entomologist who had spent three years in England with the pioneer biometrician R. A. Fisher (81), and had applied the statistical approach so as to compare the killing power of various insecticides. Bliss helped the FDA biochemists to formulate protocols for both acute and chronic toxicity studies of the glycols in various animals (78). The first paper, by Laug, Calvery, Herman Morris, and Geoffrey Woodard, appeared in 1939 (82). Other articles followed (24, 83, 84, 85).

The results presented in these papers, and the methodology underlying the results, came at a crucial time, the initial stage of the chemotherapeutic revolution, during which hundreds of new drugs would enter the medical marketplace. Elixir Sulfanilamide had waved the warning flag of toxicity, and in response FDA scientists had developed sophisticated methods for studying the toxicity of chemicals in animals. Also, as to structure, the Division of Pharmacology served as a prototype research group, imitated by universities and by the most forward-looking elements of the pharmaceutical industry, which sent scientists to observe how the FDA team went about its mission and then formed or reformed research groups of their own (78).

Influence on Food and Drug Law

Inadequacies in the Food and Drugs Act of 1906, recognized by regulators from the start, had been frequently voiced in public. The Republican ascendancy of the 1920s, however, did not provide a climate conducive to reform (53, 86, 87). At the very start of the New Deal, under the aegis of Rexford G. Tugwell, assistant secretary of agriculture, a complete revision of the law was introduced into the Congress. In a host of ways this measure would have toughened regulation of the food, drug, and cosmetic industries. Had the bill become law, it might have averted the Elixir Sulfanilamide disaster, for it would have licensed drug manufacturers.

All elements of industry combined to condemn the "Tugwell bill" which, it was declared, would turn the FDA into a sinister machine to sovietize American manufacturing. The bill had been introduced into the Senate by Royal S. Copeland, a conservative homeopathic physician from New York, who began a long campaign of compromise aimed at getting the bill into a form that might make it acceptable to members of the Congress. President Franklin D. Roosevelt gave food and drug law reform a very low priority on his list of New Deal objectives.

In October 1937, when Elixir Sulfanilamide hit the headlines, the Congress was in recess (53). A food and drug bill, much diluted from its 1933 form, had passed the Senate but was bottled up in a committee of the House. Walter G. Campbell, FDA's chief, in his first public announcement about Elixir Sulfanilamide, saying that the agency was assigning its entire field force to track down the drug, linked the disaster to the languishing bill (18). Other poisonous drugs, which the FDA could not keep off the market, had already killed and injured many citizens (86). The bill pending before the Congress did not contain any provisions that would have prevented the Elixir Sulfanilamide tragedy. Public policy required, Campbell insisted, the reinvigoration and the enactment of that bill. A federal licensing system for drugs might well be required (18).

The report that the Secretary of Agriculture made to the Congress on the Elixir Sulfanilamide affair reiterated this argument (23). The report cited a letter written to President Roosevelt by the mother of one of the "elixir's" first victims in Tulsa, a little girl of six. The mother told how her child had died in agony, and begged the President to get a law so that other families would not have to suffer as hers had done. With the letter the mother sent a picture of her daughter's smiling face (88, 89).

The "simple, direct, straightforward licensing authority" which the FDA hoped to be given as a result of the Elixir Sulfanilamide pressure proved to be too extreme a remedy for Congress to grant (90, 91, 92). Nor would Congress consider continuous inspection of drug manufacturing, analogous to the pattern prescribed in 1906 for meat processing. Industry opposition was far too strong. Some industry counter-suggestions would have eliminated any type of pre-marketing controls. Senator Copeland again fashioned a compromise, based on but stronger than a suggestion by the Proprietary Association's counsel (93), which Congress accepted and eventually included, as section 505, in the Food, Drug, and Cosmetic Act (53).

These results did not come as quickly as a consequence of the Elixir Sulfanilamide pressure as might have been expected (53, 94). From 1933 to 1937, amidst the depression and intrigued by more publicized New Deal battles, the general public knew little about the protracted Congressional struggle to secure a new food and drug law. This contest had made the front page of the *New York Times* only once, when a fight broke out in the Senate gallery (94).

Elixir Sulfanilamide did bring "a new emotional content" and greater urgency to the consideration of the food and drug bill (53). The press headlined the crisis and linked the disaster to the unenacted law. Letters from constituents to members of the Congress increased in volume. Committed partisans of reform became more vigorous and outspoken, employing sharper language. Segments of industry, opposed to the 1933 draft but satisfied by subsequent compromise, urged an end of delay. "It is a fine thing," editorialized *Advertising and Selling*, "to take a firm stand against sin and sulphanilamide . . . but . . . that isn't good enough" (95). Business should approve the whole bill. Even the most intransigent foes of legislation were somewhat shaken.

Nonetheless, more than eight months elapsed between the first reports of deaths from Elixir Sulfanilamide and the enactment of the law (53). Two more major obstacles had to be overcome--who should control food and drug advertising, the FDA or the Federal Trade Commission, and what process of court appeal should govern FDA regulation-making--before enough consensus came to get the law passed.

Licensing may come sometime, Campbell wrote after the new law had gone into effect, but the climate was not right for it in 1938 (90). "Meanwhile, there seems no alternative . . . except to undertake, through the gradual development of powers that have been conferred upon us by . . . section 505 and the authority to make factory inspections, a more effective regulation of the production and marketing of pharmaceutical products than was possible under the old statute. . . ."

Section 505--the "new drugs" provision--defined a new drug as one "not generally recognized among experts qualified by scientific training and experience to evaluate the safety of drugs, as safe for use under the conditions" listed in the labeling (96). The law forbade the marketing of new drugs in interstate commerce until their manufacturers had persuaded FDA officials that the drugs were safe. The application submitted for the agency to consider had to include samples of the drug, a list of its components and a statement of its composition, a description of the methods of manufacture, and full reports of investigations made to determine whether or not the drug was safe for use. Should the FDA raise no protest within a specified time limit, the drug could be released in the marketplace. If the agency, for cause, refused the application, the drug was barred from interstate commerce. Rights of appeal from FDA's decisions to the federal courts were specified. The law permitted the use of drugs for investigational purposes by qualified experts in order to test for safety.

This "new drug" provision, by holding pharmaceutical companies to stricter standards, also spurred them toward innovation. Competent scientists added to industrial staffs to ensure the safety of some new drugs also spent time exploring what other new drugs might be discovered. In the early 1940s a commentator credited the 1938 law with having "encouraged research" (97). This added more charge to the explosion of chemotherapy.

A minor precedent for premarket testing of chemicals in the 1938 law had been established, on a voluntary basis, shortly after the 1906 law had gone into effect, with respect to coal-tar dyes used to color foods (98). The new drugs clause of 1938, in its turn, became a more significant precedent for later laws requiring the establishment of safety before the release of pesticide chemicals (1954), food additives (1958), color additives (1960), and medical devices (1976) (99, 100). In 1962, by the Kefauver-Harris Amendments, the Congress added the requirement that proof of efficacy be demonstrated before a new drug could be released.

Research in chemistry which aimed at practical applications burgeoned in the years following World War I. Two strands of this vigorous endeavor--the commercialization of glycols and the quest for chemotherapeutic agents--met in tragic conjunction in 1937 in the Elixir Sulfanilamide affair. From the tragedy, however, and from the fear and outrage stimulated by it, emerged a system of controls under which new drugs and new chemicals affecting foods are regulated in the United States. These events helped move a complex bill, which after five years of consideration seemed moribund, to enactment by the Congress. The climate that prevented a more forthright and rigorous mode of control, outright licensing of pharmaceutical manufacturers, has not changed, as Walter Campbell then thought in time it might

(90). Indeed, in recent decades, critics of the regulatory system have argued that some relaxation in the rigor of new drug regulation might serve the public interest, thus precipitating a heated new debate in an area of continuing crucial concern to the American public (100, 101).

Literature Cited

1. S. E. Massengill Co. Massengill Briefs. No. 1, Oct. 1937, in Administrative File 1258, Food and Drug Administration Records, Record Group 88, Washington National Records Center, Suitland, MD.
2. Dowling, Harry F. "Fighting Infection"; Harvard: Cambridge, 1977, Chapter 8.
3. Silverman, Milton. "Magic in a Bottle"; Appleton-Century: New York, 1943, Chapter 10.
4. Domagk, G. Deutsche Med. Wchnschr. 1935, 61, 250-3.
5. Obituary. Lancet. 1964, 1, 992-3.
6. Hare, Ronald. "The Birth of Penicillin and the Disarming of Microbes"; George Allen and Unwin: London, 1970, Chapters 7 and 8.
7. Colebrook, Leonard; Kenny, Meave. Lancet. 1936, 1, 1279-86.
8. Colebrook, Leonard; Kenny, Meave. Lancet. 1936, 2, 1319-22.
9. Colebrook, Leonard; Buttle, G. A. H. Lancet. 1936, 2, 1323-6.
10. Long, Perrin H.; Bliss, Eleanor A. J.A.M.A. 1937, 108, 32-7.
11. Long, Perrin H.; Bliss, Eleanor A. South. M. J. 1937, 30, 479-87.
12. N. Y. Times. Dec. 17, 1936, p 1.
13. N. Y. Times. Dec. 18, 1936, p 24.
14. N. Y. Times. June 25, 1937, p 16.
15. N. Y. Times. July 1, 1937, p 1.
16. Nelson, E. E. Food Drug Cosmetic Law J. 1951, 6, 344-53.
17. A. G. Murray, FDA senior chemist, to District chiefs, Nov. 8, 1937, AF 1258.
18. Baltimore Sun, Oct. 20, 1937, clipping, AF 1258.
19. Newman, B. A.; Sharlit, H. J.A.M.A. 1937, 109, 1036-7.
20. Perrin H. Long to Herbert O. Calvery, Dec. 7, 1937, decimal file 510-.20S for 1938, FDA Records, WNRC.
21. Theodore G. Klumpp, chief medical officer, FDA, report on Massengill Co., Oct. 18, 1937, AF 1258.
22. Theodore G. Klumpp, report on Oct. 18, 1937, visit to Massengill Co., [Apr. 1938], AF 1258.
23. Wallace, Henry A. "Report of the Secretary of Agriculture on Deaths Due to Elixir-Sulfanilamide-Massengill." 75 Cong. 2 sess., Senate Document 124, Serial 10247.
24. Curme, George O.; Johnston, Franklin. "Glycols"; American Chemical Society Monograph Series; Reinhold: New York, 1952.

25. Report of the Eastern District for 1937-1938, July 26, 1938, decimal file .053 for 1938, FDA Records, WNRC.
26. A. W. Lowe to chief, Central District, Oct. 27, 1937, AF 1258.
27. von Oettingen, W. F.; Jirouch, E. A. J. Pharmacol. & Exper. Therap. 1931, 42, 355-72.
28. Haag, H. B.; Ambrose, A. M. J. Pharmacol. & Exper. Therap. 1937, 59, 93-100.
29. Editorial. J.A.M.A. 1937, 109, 1367.
30. Editorial. J.A.M.A. 1937, 109, 1456.
31. Burrow, James G. "AMA: Voice of American Medicine"; Johns Hopkins: Baltimore, 1963, p 74-5, 126.
32. AMA Chemical Laboratory. J.A.M.A. 1937, 109, 1530-9.
33. Walter E. Donaldson to chief, Central District, Oct. 15, 17, 29, 1937, AF 1258.
34. S. E. Massengill Co. "S. E. Massengill's Experience under Federal Food and Drugs Act"; S. E. Massengill Co.: Bristol, [July 1938], in AF 1258.
35. Massengill, S. E. "A Sketch of Medicine and Pharmacy and a View of Its Progress by the Massengill Company"; S. E. Massengill Co.: Bristol, 1943, p. 209, 233, 420, 424, 428.
36. Carl S. McKellogg report, Aug. 22, 1941, AF 1258.
37. Anon. Time. Dec. 20, 1937, 30, 48-9.
38. Bristol Herald Courier, Dec. 2 1937, clipping, AF 1258.
39. Notices of Judgment under the Food and Drugs Act, 23228 (1935).
40. Notices of Judgment under the Food and Drugs Act, 27136 (1937).
41. Notices of Judgment under the Food and Drugs Act, 24029 (1935).
42. George Larrick to chief, Central District, Nov. 21, 1934, AF 1258.
43. J. J. McManus to chief, Central District, Nov. 11, 1937, AF 1258.
44. Anon. Food and Drug Rev. 1949, 33, 117.
45. Anon. Food and Drug Rev. 1937, 21, 267-9.
46. Young, J. H. Emory University Qtly. 1958, 14, 230-7.
47. McManus, J. J. Food Drug Cosmetic Law J. 1956, 11, 194-5.
48. Young J. H.; Janssen, W. F. Oral History of the U. S. Food and Drug Administration, with John J. McManus and Clarence D. Schiffman (1968), History of Medicine Division, National Library of Medicine, Bethesda, MD.
49. J. J. McManus to chief, Central District, Nov. 9, 1937, AF 1258.
50. Chronological file of Inspectors' and others' reports. AF 1258.
51. Central District report on enforcement operations, 1938, decimal file .053 for 1938, FDA Records, WNRC.
52. Campbell, W. G. "1938 Report of Food and Drug Administration"; FDA: Washington, 1938, p 13.

53. Jackson, Charles O. "Food and Drug Legislation in the New Deal"; Princeton University Press: Princeton, 1970, Chapter 7.
54. Notices of Judgment under the Food and Drugs Act, 29751 and 29752 (1939).
55. Paul B. Dunbar to Drug Division, July 23, 1938, AF 1258.
56. A Description of Glycols (1937), AF 1258.
57. The Facts about Elixir Sulfanilamide (1937), AF 1258.
58. W. T. Ford to chief, Cincinnati Station, July 21, 1938, AF 1258.
59. Walter G. Campbell to Burrow & Burrow, Jan. 21, 1938, AF 1258.
60. Correspondence in decimal file 510-.20S for 1938, FDA Records, WNRC.
61. Walter G. Campbell to Solicitor Mastin G. White, Feb. 17, 1938, AF 1258.
62. Walter G. Campbell to Paul H. Appleby, assistant to the Secretary of Agriculture, June 23, 1938, Food Law folder for 1938, Records of the Office of the Secretary of Agriculture, Record Group 16, National Archives and Records Service, Washington, DC.
63. Paul B. Dunbar to Solicitor, Aug. 31, 1938, AF 1258.
64. Paul B. Dunbar to Secretary of Agriculture, Sep. 7, 1938, AF 1258.
65. Anon. Food and Drug Rev. 1938, 22, 230.
66. Paul B. Dunbar to J. B. Frazier, Jr., U. S. Attorney in Knoxvile, Oct. 5, 1938, AF 1258.
67. Crawford, Kenneth G. "The Pressure Boys"; J. Messner; New York, 1939, p 73.
68. Cowen, David L., in Blake, John B., Ed.; "Safeguarding the Public: Historical Aspects of Medicinal Drug Control"; Johns Hopkins, Baltimore, 1970, p 72-82.
69. Weikel, Karl. Revised master's thesis. American Institute of the History of Pharmacy, University of Wisconsin, 1963.
70. Mahoney, Tom. "The Merchants of Life"; Harper, New York, 1959, p 4.
71. Merck, George W. "Extracts from Welcoming Address"; Merck & Co.: Rahway, 1933.
72. Young, J. H., in Garraty, John A., Ed.; "Dictionary of American Biography, Supplement Six"; Scribner's: New York, 1980, p 447-8.
73. Editorial. J.A.M.A. 1937, 109, 1128.
74. Denkewalter, R. G.; Tishler, Max, in Jucker, E., Ed.; "Fortschritte der Arneimittelforschung"; vol. 10, Birkhauser: Basel, 1966, p 11-31.
75. Anon. Food and Drug Rev. 1938, 22 p 16, 26.
76. J. O. Clarke to chief, Cincinnati District Station, Jan. 3, 1938, decimal file 004.1 for 1938, FDA Records, WNRC.

77. Hile, Joseph P. "Teddy Makes It All Possible: A Historical Review of FDA"; address at Association of Food and Drug Officials, June 1981.
78. Young, J. H.; Lofsvold, F. L.; Janssen, W. F.; Porter, R. G. Oral History of the U. S. Food and Drug Administration: Pharmacology, with Vos, B. J.; Fitzhugh, O. G.; Laug, E. P.; Woodard, G. (1980), History of Medicine Division, National Library of Medicine.
79. Trevan, J. W. Proc. Roy. Soc. London s.B (Biol.) 1927, 101, 483-514.
80. Anon., in Debus, Allen G., Ed.; "World's Who's Who in Science"; Marquis, Chicago, 1968, p 191.
81. Gridgeman, N. T., in Gillespie, Charles C., Ed.; "Dictionary of Scientific Biography"; vol. 5, Scribner's: New York, 1972, p 7-11.
82. Laug, E. P.; Calvery, H. O.; Morris, H. J.; Woodard, G. J. Indust. Hyg. and Toxicol. 1939, 21, 173-201.
83. Calvery, H. O.; Klumpp, T. G. South. M. J. 1939, 32, 1105-9.
84. Morris, H. J.; Nelson, A. A.; Calvery, H. O. J. Pharmacol. & Exper. Therap. 1942, 74, 266-73.
85. Fitzhugh, O. G.; Nelson, A. A. J. Indust. Hyg. and Toxicol. 1946, 28, 40-51.
86. Lamb, Ruth D. "American Chamber of Horrors"; Farrar and Rinehart: New York, 1936, Chapter 4.
87. Young, James H. "The Medical Messiahs"; Princeton University Press: Princeton, 1967, Chapter 8.
88. Secretary of Agriculture's Report to the Senate (carbon copy of original), Bills--Res. H. 352 file, 1937, Records of the Office of the Secretary of Agriculture, NARS.
89. Krieghbaum, Hillier. Survey Graphic, 1938, 27, 271-4.
90. Walter G. Campbell to Arthur E. Paul, May 23, 1941, decimal file 062.1 for 1941, FDA Records, WNRC.
91. Henry A. Wallace to A. T. McCormack, Nov. 6, 1937, Drugs file for 1937, Records of the Office of the Secretary of Agriculture, NARS.
92. Walter G. Campbell to Royal S. Copeland, Feb. 7, 1938, decimal file 062 for 1938, FDA Records, WNRC.
93. Ruth Lamb to Robert Littell, Feb. 4, 1938, decimal file 062 for 1938, FDA Records, WNRC.
94. Cavers, D. F. Law and Contemporary Problems. 1939, 6, 2-42.
95. Editorial, Advertising and Selling, Nov. 18, 1937, 30 11.
96. 52 U. S. Stat. 1040.
97. Anderson, C. M. Food Drug Cosmetic Law Qtly. 1946, 1, 71-85.
98. Hochheiser, Sheldon, chapter in this book.
99. Janssen, W. F. J. Public Law. 1964, 13, 205-21.
100. Temin, Peter, "Taking Your Medicine: Drug Regulation in the United States"; Harvard: Cambridge, 1980.
101. Young, J. H. Pharmacy in History. 1982, 24, 3-31.

RECEIVED November 10, 1982

7

The Establishment of Synthetic Food Color Regulation in the United States, 1906–1912

SHELDON HOCHHEISER
Rohm and Haas Company Philadelphia, PA 19105

In the years immediately following the 1906 enactment of the first federal food and drug act, Dr. Bernhard Hesse made a thorough study for the government of synthetic food colors in the marketplace and from this study developed a regulatory scheme based on two principles: 1) only a short explicit list of colors should be permitted; 2) each batch of these colors manufactured should be certified by the government as to its identity, purity, and freedom from harmful substances. After substantial internal disagreement, Department of Agriculture officials concluded that there was no statutory provision authorizing this scheme, and adopted a plan based on voluntary participation. This plan proved successful in providing the country with an ample supply of safe food colors.

In the last third of the nineteenth century, the United States food supply system underwent a major transformation, the most fundamental one in the history of the nation. In 1860, most Americans lived on farms and largely ate what they produced. Even in urban areas the foods for sale were chiefly raw, unprocessed, and of local origin. By 1900, America was substantially urbanized, and much of the food sold in the cities was processed--refrigerated, canned, chemically preserved--by large firms doing business on a national scale. With this growing distance between producer and consumer came an increasing opportunity for artifice in the food supply. The unannounced substitution of a cheaper food for a more expensive one and the use of added substances to preserve and color, although practiced since antiquity, became more widespread, and new techniques, such as the use of synthetic colors prepared from coal tar, became common (1).

Coal tar dyes first came on the market in the late 1850's, and within a few years had virtually replaced natural dyestuffs in the textile trade. Almost from the first, these same dyes

0097–6156/83/0228–0127 $06.00/0

were sold for the less voluminous food color trade, where they were seen as a great improvement over the often acutely toxic mineral pigments they, in many cases, replaced. By the mid 1870's, the entire dye industry was controlled by a cartel of German firms, who through a combination of research support and aggressive capitalism, controlled the sale of both the intermediate chemicals and the dyestuffs themselves (2).

One result of the transformation of the food system was a growing recognition of the need for the government to regulate the food supply, both to protect the processor from unfair competition, and to protect the consumer from fraud or harm. The fight for federal legislation to provide for such regulation, led by Harvey Wiley, chief of the Bureau of Chemistry of the United States Department of Agriculture, lasted over twenty years. It concluded with the enactment of the Pure Food and Drug Act of 1906, commonly known as the Wiley Act (3).

To Wiley, the Food and Drug Act was his mandate to insure the country a pure and honest food supply more than it was a specific statute with specific legal requirements and limitations. Thus, although the law mentioned colors only to prohibit their addition to conceal inferiority and to ban their use, if poisonous, in confectionary, Wiley believed, as part of his overall suspicions of chemical additives, that the use of coal tar colors was a practice requiring, at the least, thorough investigation. As this would require time, Wiley arranged for color policy to be deferred from the initial regulations that were to be issued before the act became effective on January 1. When the Department of Agriculture issued the regulations in October, it noted simply that "the Secretary of Agriculture shall determine from time to time . . . the principles which shall guide the use of colors" (4). Neither Wiley nor any of his staff was familiar with these dyes, so in mid summer he hired an outside consultant, Dr. Bernhard C. Hesse of New York City, to study the question. Hesse agreed to devote twenty hours a week to this project (5).

Hesse was unusually, if not uniquely, qualified to examine the coal tar dye question for the United States government; he was an American expert in a German dominated field and industry. Hesse was born in East Saginaw, Michigan in 1869. He earned degrees in both pharmacy and chemistry from the University of Michigan, and in 1896 became the third recipient of the Ph.D. in Chemistry at the new University of Chicago. Upon graduation, Hesse took a position with Badische Anilin ünd Soda Fabrik. Badische initially sent him to their headquarters in Germany, but soon returned him to the United States where, based in New York, he served as Badische's resident technical expert in patent litigation until the end of 1905, when he began an independent consulting practice. Thus Hesse had been privy not only to many of Badische's trade secrets, but also to matters of company and cartel policy. So when Hesse accepted the assignment from Wiley, he brought to the service of his country years of experience with the dominant foreign dye interests (6,7).

Initially, Hesse expected to complete within three months what he saw as a straightforward exercise. As he wrote Wiley in August of 1906, the primary task appeared to be gathering and evaluating representative literature, opinions, and samples and devising a scheme for the chemical separation and detection of syestuffs in foods (8).

By November, he was finding the task somewhat more complex. Since returning to New York in September, he had collected some ninety different samples of dyes (some of them duplicates) recommended by their distributors for use in food, a far greater number than he had expected. Some of the colors so recommended, such as Orange II, Hesse knew to be far from harmless, even if technically suitable. The poor quality and high level of impurities he found in the samples surprised him. He now thought it would be necessary, in addition to the tasks he had outlined in August, to consider "the question of inspection of coloring matters suitable for use in food products" (9). This question appeared to be rather complicated. Wiley agreed that the situation looked far more complex than it had originally seemed, and told Hesse to proceed and take the time necessary to do a thorough job (10).

Proceed Hesse did, but he confessed to Wiley in February that he was still far from finished. The analytical separation and detection work was in an early stage. He had not completed methods for the various red dyes, and had not started any other color group. The investigation would have to be redirected, and the analytical portion put aside until after the development of definite regulations. "The aim of this work ought to be to get a fair 'standard of purity' for each chemical individual" and to hold each manufacturer to this standard. It might prove necessary to permit the sale of pure colors only, with mixing and dilution to be done at the point of use. In a few months, Hesse's task had evolved from analysis and literature search to the devising of regulations (11).

While Hesse, in New York, was discovering all these difficulties, Wiley, in Washington, was encountering far greater problems of his own. The crusading zeal that had served him well in the long struggle for the passage of a food and drug law brought him into ever increasing conflict with his superior, Secretary of Agriculture James Wilson. Wiley fought with Wilson on several issues--on the definition of blended whisky, on the use of sulfur dioxide in the preservation of dried fruit, on the need to seek advice from the department's legal staff.

By March 1907, Wilson had concluded that he, who after all was legally responsible for the law, could no longer rely exclusively on Wiley, whom he concluded tended to take arbitrary theoretical positions that unnecessarily irritated business and agriculture. In April Wilson hired another chemist, Dr. Frederick Dunlap, to provide him with an independent opinion on the often complex scientific points affecting food and drug regulation, and established a three member Board of Food and Drug Inspection consisting of Wiley (as chairman), Dunlap and George McCabe, the Department Solictor. The Board's job would be to consider all questions on regulation, enforcement, and violations of the food and drug law, to advise the Secretary, and to prepare, for his signature, Food Inspection Decisions (FIDs), published regulations and interpretations. Wiley was outraged, especially as Wilson had announced his decision by arriving at Wiley's office one morning with Dunlap in tow (12).

The Board, presumably at Wilson's urging, was anxious to get the color question (and other pending matters) settled. Thus, while Wiley had urged Hesse to take his time and be thorough, the Board as a whole pressed him for an early report. Hesse replied that while he felt nowhere near finished, it might be possible, as an interim measure, to issue a short list of colors that would be considered harmless and permitted for use. For the first time, he was proposing restricting food dyes to a short explicit list, one that would contain only harmless colors, and as few of those as necessary to provide an adequate variety of shades. The list he proposed contained seven colors, two yellows (napthol yellow S and tartrazine), one green (light green SF yellowish), and four reds (carmoisine, rhodamine B, amaranth, and erythrosine). Hesse chose these dyes by taking the ten dyes that had been recommended by three of the five suppliers from whom he had collected samples to date, and then eliminating three for which he found an unfavorable toxicological evaluation in one of three sources (13).

The Use of Coal Tar Colors in Food Products by Hugo Lieber was the first such source. Lieber, an importer and merchant of coal-tar colors, privately published his book as a promotional device. The toxicological studies Lieber reported were actually taken from a highly regarded 1888 book by the German physiological chemist Theodor Weyl. Lieber barely credited Weyl in a single textual reference. It seems curious that Lieber distributed this abridged version, or that Hesse used it, as a complete translation of Weyl's book had been published in Philadelphia in 1892. Weyl reported on animal experiments he and others had performed with a number of coal tar colors (14,15).

Hesse's second source, Die Arzneimittel Synthese by Sigmund Fränkel, was a largely theoretical work by a prominent scientist. Fränkel tried to establish direct and specific relationships between chemical structure and physiological activity. In this book Hesse found broad assertions, such as the statement that all unsulfonated azo dyes were harmful, as well as specific claims. Although it was perhaps natural that, as an organic chemist, Hesse would find Fränkel's ideas persuasive, they were far from unanimously accepted by pharmacologists (16,17). Hesse's final source was a list of colors, apparently a trade circular, prepared in 1899 by the National Confectioners Association.

Dunlap and McCabe wanted a full report from Hesse, not merely a letter, because they hoped to prepare at least an interim FID on coal-tar colors for Wilson's early approval. In May, they arranged for Hesse to appear before the Board in Washington on June 10, 1907.

Hesse spent the intervening month putting his data and notes in order. He made no attempt to salvage his analytical detection work. It was both too far from completion and not of immediate concern. Weyl's book replaced Lieber's as his primary source for toxicological data. He visited additional dye suppliers, including the subsidiaries of two big German firms (Badische and Hoechst), American agents for several smaller European companies, two domestic manufacturers of coal-tar colors (Schoelkopf, Hartford, and Hanna of Buffalo and Heller, Merz and Company of Newark), and two smaller American businesses which specialized in the importation and packaging of colors for the food trade (H. Kohnstamm and Company and H. Lieber and Company, both of New York). By June 10, Hesse had 203 samples from eight firms, of which 109, including 53 different chemicals, came with definite chemical identification in the form of their table number in the standard reference work generally known as the Green tables (18). Applying the principles he had enunciated in May, he tabulated these 53 colors by the number of firms offering them, and discovered that only eleven colors were offered by as many as half of his respondents. He rejected four of these on the basis of unfavorable reports by Weyl, leaving a list of seven colors, five reds and two yellows.

At his meeting with the Board, Hesse explained his work to date and the sorry, confused state of the small portion of the coal-tar color trade devoted to food. With the single exception of the specialty house of Kohnstamm, the industry refused to follow or guarantee their products into the food trade even as they competed with advice and guarantees in the textile trade. A single dye often would be available in widely varying levels of purity. Methylene blue, the only coal-tar color in the United States Pharmacopeia, was an extreme example. It sold at prices from \$75 a pound when prepared for human injection, and \$21.50 when prepared to USP standards for ingestion, down to \$1.25 for textiles. This showed that the industry was capable of producing highly purified chemicals when it was convinced it was necessary and profitable to do so. Hesse saw no evidence of any manufacturer taking this sort of care with the dyes it sold for food use. He was unable to find agreement even on the levels of dye that were commonly used--one consulting chemist told him 1 part in 3,000 was not uncommon in candy, while another said that no more than 1 part in 50,000 was advisable.

Hesse complained that even if one could get an industrial representative to speak of harmless coal tar dyes, he could not be trusted to give an honest answer. A Dr. Schweitzer of the Elberfeld company gave Hesse a list of some forty-three colors that he asserted were harmless in food products. He cited Weyl as his authority for fifteen of these but Hesse could find no reference to six of these colors in Weyl's book, and only negative comments on three more.

Finally, Hesse gave the Board his "provisional conclusions for today" and proposed that the government regulate food colors by permitting the use of only a small list of expressly permitted colors while prohibiting the use of all others. He suggested that the Board issue a tentative list of seven permitted colors, the two yellows and four reds from his previous list, and light green SF bluish (Green table #434), without giving any reasons for his selection of these, but with a request for time for laboratory examination of commercial samples of the seven colors in order to determine the level of purity thereof (19).

Hesse indeed presented the above list as his conclusion for that specific day, for only five days later he wrote to Wiley, as chairman of the Board, substantially revising it. He decided that the composition of the list could best be determined by a sort of popularity contest with the selection as single allowed representative of a given shade the chemical with a clean toxicological record that had been submitted by the largest number of firms. In this manner, he culled from his samples a list of fifteen colors, representing the top vote getter in each of fourteen different shades (two oranges tied with two votes each). He eliminated six of the fifteen on the basis of unfavorable toxicological reports, although, curiously, he did not look to see if another less popular color could take their places. This left nine shades. He eliminated two more because he felt others in the list could replace them when used in combination, leaving once again seven colors, as follows:

Table I: Hesse's recommended list of 6/15/07

Shade	Color	Green Table No.
Red	Amaranth	107
Scarlet	Panceau 3R	56
Bluish Red	Erythrosine	517
Orange	Orange I	85
Yellow	Napthol Yellow S	4
Green	Light Green SF Yellowish	435
Blue	Indigo Disulphonic Acid	692

Thus, Hesse replaced four of the seven colors he had recommended only five days before, solely on the grounds that another chemical of similar shade had been submitted, and thus desired, by more sources. He implicitly assumed either that all colors of a given shade were of equal technical suitability for food use or that suitability was in direct proportion to the number of firms offering the shade. This tenet, he would eventually be forced to concede, was invalid in at least one case.

He concluded that:

"My tentative ruling of last Monday ought to be put in this final form subject to restriction as further investigations warrent . . .(These seven dyes) may for the present be used as food colors. Each of these colors shall be free from any coloring-matter other than the one specified, shall not contain any contaminations of imperfect or incomplete manufacture, and these colors when mixed with one another for shading purposes shall be so mixed at the place and at the time of use and not otherwise nor elsewhere" (20).

The Board was so eager to send Secretary Wilson a decision that it unanimously accepted Hesse's recommendations, except for the section on mixing, at its next meeting on June 17, 1907, and made it part of Food Inspection Decision 76, "Dyes, Chemicals, and Preservatives," which was issued over the signature of the three Board members, and the Secretaries of Agriculture, Treasury, and Commerce and Labor on June 18, 1907. In printing Hesse's recommendations, the Board explained its attitude toward the use of coal tar colors:

"The use of any dye, harmless or otherwise, to color or stain a food in a manner whereby damage or inferiority is concealed is specifically prohibited by law. The use in food for any purpose of any mineral dye or any coal-tar dye, except those coal-tar tyes hereinafter listed, will be grounds for prosecution. Pending further investigations now underway and the announcement thereof, the coal-tar dyes hereinafter named, made specifically for use in foods, and which bear a guarantee from the manufacturer that they are free from subsidiary products and represent the actual substance the name of which they bear, may be used in foods. In every case a certificate that the dye in question has been tested by competent experts and found to be free from harmful constituents must be filed with the Secretary of Agriculture and approved by him" (21).

With this the Board established the basic principles of food color regulation in the United States. The permitted dyes were those on Hesse's list. Parts of this decision would soon become open to legal question.

A supplementary memorandum from the Board accompanied FID 76. It discussed Hesse's background, qualifications, plan of work and the methods he used in reaching his recommendations. It explained the composition and small size of the permitted list as providing a range of unpatented colors versatile enough to provide for all legitimate uses while allowing only colors with clean toxicological records. It justified the requirement for a manufacturer's guarantee by noting, as Hesse had told Wiley, that it was the manufacturer who should know the nature of his goods and that they were as represented and free from harmful impurities (22).

Hesse next went to the color industry to find its reaction to this program and to translate the requirement for purity into definite standards which would be stringent, but no more so that the best achievable by good manufacturing practice. Among others, he met with E. G. Kohnstamm, the President of H. Kohnstamm and Company. Kohnstamm found the regulations generally reasonable and workable, agreeing the seven colors could serve for all purposes, with mixtures preparable to any desired shade. But then, Kohnstamm's was the only firm which had previously expressed a willingness to guarantee colors. He naturally had questions on the yet unannounced form of the manufacturer's guarantee, but did not see it as a problem and also wondered what tests would suffice to demonstrate the necessary levels of purity. Hesse would not commit himself to any specific test, as he had not developed his ideas on the subject fully, and he hoped to learn if the manufacturers had any tests in use. The two men agreed that relevant tests and standards from the *United States Pharmacopeia* might be a good starting point for these tests.

On one point Hesse and Kohnstamm disagreed. Kohnstamm interpreted FID 76 as permitting the sale of mixtures of certified colors, giving Solicitor McCabe as his source for this interpretation. Kohnstamm had always sold his food colors as tradenamed mixtures of standardized shade and tinctorial strength, and feared that requiring the sale of pure colors only would simplify the business to such a degree that most firms, including his own, would be driven from the business. McCabe's interpretation disappointed Hesse as it was contrary to the chemist's recommendation. At this time, Hesse would have welcomed the oligopolization of the trade by a small number of firms, as he expected would happen if sale of mixtures were banned. He thought that the largest firms could best produce and maintain high quality goods (23).

On July 31, the Board formally rejected Hesse's recommendation on mixtures. This was yet another indication of a growing rift between Wiley's hard line attitude and Wilson and McCabe's more cautious approach (24).

In August, Hesse began to realize that his long standing faith in the large German firms had been mistaken. He left an early August meeting with two representatives of Badische, his old employer, quite disgusted. These men showed little willingness to work within the regulations. Rather, they imperiously demanded the listing of all their colors, and a simple form that they could fill out once to permanently register their products. They objected when Hesse told them that each batch might require separate tests and guarantee and the implication from this that a manufacturer would be continuously responsible for the harmlessness of the goods it sold.

At the very least, Badische wanted the government to provide it with a set of tests to be used. Perhaps this last objection bothered Hesse the most, for it indicated to him that Badische, and thus probably the entire industry, did not test dyes intended for food use for freedom from hazardous impurities and therefore knew of no such tests (25). An agent for another manufacturer went further, telling Hesse that no distinction was ever made between food colors and textile colors in his warehouse: "'Napthol Yellow S for wool and Napthol Yellow S for food all comes out of the same barrel'" (26). At best, a manufacturer guessed that a dye was suitable. Hesse suspected that the quality and variety of the product that the Bureau would receive would demonstrate this to be the case, and thereby justify any arbitrariness by the government (25).

Badische decided to withdraw from direct participation in the American food color trade rather than submit to the requirements of FID 76. It opted instead to sell the dyes to middlemen such as Kohnstamm who could, after further purifying the dyes if desired, submit the guarantees and assume legal liability. The other German firms soon followed Badische's lead.

Kohstamm began doing as Badische suggested. It bought available grades of the seven colors from Badische and prepared to purify further, test, and submit the dyes for government approval itself. Where the standard grades seemed not good enough, it pressed Badische for superior goods. Hesse decided to wait and see what tests and standards companies like Kohnstamm would develop and to use these to push the submitters into producing the cleanest possible goods (27).

In FID 76, the Board had left open the details of the certificates a manufacturer would need to submit to obtain approval for his goods. Kohnstamm and several others asked Hesse or Wiley what form these should take. The Board responded to these inquiries in September by issuing FID 77 which established legal procedures for certification, made clear the requirement that each batch of dye be separately certified, and defined several terms. Forms were provided for four different certificates: two ("foundation" and "manufacturer's") for submission with an initial batch of a color, one ("supplementary") for additional batches, and one ("secondary") for mixtures and repacks. FID 77 defined "batch" as a single lot of dye processed in a single manner at a single time, "mixture" as a blend of two or more certified colors and "competent expert" as anyone who through training or experience could devise and perform the necessary tests for identity and purity (28).

Although the Board had been receiving, and rejecting as insufficient, manufacturer's guarantees since June, large numbers of foundation and manufacturer's certificates began reaching the Board in the fall, demonstrating a substantial demand for certified colors. A number of food manufacturers wrote to Hesse or Wiley requesting sources of certified colors, and several food ingredient jobbers distributed circulars to the food trade offering for sale "the colors mentioned in FID 76", falsely implying, at least to Hesse, that they had certified colors for sale. Much of the food and color industries accepted FID 76 as mandating the use of certified colors only (29).

The Board forwarded each of the several hundred foundation

certificates it received to Hesse for comment and criticism. Hesse found the overall quality of both the tests and the dyes themselves depressingly poor, and for several months rejected all he saw. Most of the dyes submitted were no more than simple repackings of German dyes, and contained substantial impurities in direct violation of the regulations. Percentages of simple inorganic salts such as NaCl and Na_2SO_4 ran as high as 45%. These salts had been used in the fabrication of the dyes both to precipitate the dye from solution and to reduce the strength of the dye to some predetermined tinctorial standard. Although these salts may have presented little direct hazard, they tended, when used to excess in precipitations, to carry substantial impurities along with the desired precipitate. A large number of samples had high levels of chemicals closely related to the intended product, including martius yellow (a known poison) in napthol yellow S and orange II in orange I. This was evidence of sloppy manufacture. High levels of known toxins, especially arsenic and less often lead, were common in disregard for the public health (30). The jobbers who submitted these dyes were, in a way, not really at fault. They were submitting the same dyes the German manufacturers sold them as suitable for food use, in many cases the same ones they had been selling for years.

Hesse justified his rejection of these dyes of customarily low quality by pointing out that several German firms were offering, at premium prices, chemically pure grades of at least two of the intermediates used in preparation of the permitted colors, paranitroanilinie and betanapthol. Thus, a high quality product was technically feasible (31).

Hesse did not find the foundation certificates any better than the dyes. Some certificates reported no tests for impurities or toxins which Hesse believed might be present. Some of the tests were inadequately sensitive or selective and many were not described in sufficient detail for duplication, if they were described at all. These certificates confirmed Hesse's fears that the industry had not been testing food dyes, and thus had no such set of tests.

Hesse wrote a critique of each certificate, giving the reasons for its rejection and suggesting where it could be improved so that the applicant could learn from his mistakes and correct them before the next submission. The following letter, rejecting two batches of ponceau 3R submitted by Charles Shaw of Baltimore, was typical:

"You don't say what tests you use for arsenic, what organic constituents you test for, [or the] preparation of materials used in carrying out these experiments. Your dyes have 27.1% and 32.3% salt. This is way too much. You must get your foreign suppliers to give you unfilled pure dyes and submit these. If you wish to get them filled or standardized with salt, do this later and resubmit the dye for secondary certification" (32).

Kohnstamm in particular took the criticism seriously and

worked diligently to prepare dyes as pure as possible and to submit the same to ever more detailed tests. Hesse tried to assist Kohnstamm, meeting several times with members of its staff and helping them to find ways to improve the quality of the products and the associated tests.

Hesse used the results reported in these certificates to begin to formulate the standards of purity which he would expect from a certifiable dye. If one certificate reported a particularly good result for, say, salt in a given dye, that would become the temporary minimum. He strove to secure for the country not some elusive threoretically pure dyes, but simply the best that industry could do if pushed, pushed as he had become convinced it had never been pushed before (33).

On November 14, 1907, Hesse found a certificate he could approve, one submitted by Kohnstamm for a small batch of light green SF yellowish. He wrote to the Board that this certificate vindicated his uncompromising attitude and showed that clean goods, cleaner than E. G. Kohnstamm himself had thought possible that spring, could be made. Hesse praised Kohnstamm for this accomplishment, citing Kohnstamm's persuation of Badische to provide it with arsenic free dye, its investment in new equipment and research, and its diligence in correcting deficiencies Hesse found in previous submissions. Hesse advised the Board that the next step should be for a Bureau staff chemist to verify the certificate by repeating the tests therein described on the dye sample submitted.

Although the Kohnstamm certificate for light green was certainly a major step, the project was far from finished. This was only one color, and a small laboratory test batch at that. Still, it now seemed certain to Hesse that all seven colors would eventually be available for sale with certification of their quality, and the promise of FID 76, that use of any dye other than one of the seven in certified form would be ground for prosecution, could be realized.

The legal authority for this promise soon came under attack. The affected industries were watching the developments in Washington, and at least one trade group, the National Confectioners Association, did not like what it saw. Its legal counsel, Thomas Lannen of Chicago, advised the members that they did not need to worry because FID 76 and FID 77, in his studied opinion, overstepped the authority granted the Department of Agriculture by the Food and Drug Act of 1906. This opinion appeared in the December 1907 issue of the American Food Journal. Lannen explained that as the law merely prohibited poisonous colors, the Board of Food and Drug Inspection had no authority to restrict colors to any short chosen list. The Board itself was without authorization in the law, but FID 77 was the worst offender as in it the Board "made the legality or illegality of the use of a color dependent on whether or not permission has been given by the government to use it instead of being dependent on the nature of the color" (34).

Hesse and Wiley apparently were unaware of this criticism, but then neither seemed overly concerned with the letter of the law, especially if it might interfere with the protection of the food supply. McCabe, however, being a lawyer instead of a scientist, was concerned with just these points, although he had not yet brought them before Wiley or Dunlap.

Several firms raised narrower objections to the color certification process. At least one contended that Hesse's standards, as demonstrated by the rejections they had received, were ruinously high (35). Others appealed for Board authorization to sell the seven permitted colors in uncertified form, at least until sufficient certified colors were available. Hesse replied to the former objections by citing Kohnstamm's light green as proof that his standards were not unreasonable, and to the latter by claiming that such permission would undermine the entire program (36). The Board, however, took no action against those firms which advertised and sold the seven colors in uncertified form, and even informally approved the use of uncertified but permitted colors in response to the several inquiries (37,38). The Board also received requests of varying degrees of seriousness for additions to the permitted list, but chose to defer consideration of these until the system was functioning smoothly for the seven colors.

The Board's chief act on colors at this time was requiring that Hesse himself perform the verification tests on Kohnstamm's light green SF yellowish. The color proved to be as good as Kohnstamm had claimed (39,40). Kohnstamm followed its success with the green with similarly acceptable batches of ponceau 3R and orange I in February, and indigo disulfonic acid, erythrosine, and amaranth in March (41). The remaining color, naphthol yellow S, proved more difficult. It differed from the poisonous martius yellow only in the presence of a single sulfonic acid group. Thus, any degree of incompleteness in the sulfonation reaction left the dye contaminated. Kohnstamm solved this problem by mid-summer. Now that small batches of all its colors had been accepted, the firm attempted operations on a commercial scale, with batches of hundreds rather than tens of pounds. It encountered additional problems; the cast iron water pipes in the factory, for example, contaminated the water supply sufficiently to cause some batches to fail the heavy metal test.

One by one Kohnstamm succeeded with the colors. By November, it had four colors in large batches for which it had sent certificates to the Board. By the new year, it was sufficiently confident of success that on January 26, 1909 it announced to the trade that it was ready to accept orders for certified colors. Initially, orders would be accepted for the colors only as compounded into Kohnstamm's well established customary color mixtures which sold under names such as "auramine" and "chocolate brown," rather than as Hesse had hoped, as pure dyes (42,43).

Although Hesse had speculated early in 1908 that Kohnstamm might well end with a monopoly in certified colors, and that this would not necessarily be a bad thing (44), in May a second firm (Schoelkopf, Hartford, and Hanna of Buffalo, New York) began serious attempts at preparation and analysis of specially purified lots of the seven colors for certification and subsequent sale through its National Aniline Company subsidiary. Apparently, Schoelkopf took the news that a small American specialty house was succeeding in producing food colors of such high purity as a direct challenge (45). Schoelkopf was by far the largest and most successful native American manufacturer of coal-tar colors, accounting for approximately 10% of domestic consumption, and this unique position was no doubt the reason it so reacted to Kohnstamm's achievements. It had a proud record of successful innovation in the face of the German cartel, having introduced several new dyes to the textile trade, and having made great strides towards achieving its own supplies of intermediates (46).

Once Hesse, and through him Wiley, became convinced of Schoelkopf's sincerity, they began assisting the company in meeting its goal, much as they had helped Kohnstamm. Hesse held several meetings with representatives of the firm in New York and traveled twice, with Wiley's approval, to Buffalo. Once Schoelkopf started it made impressive chemical progress; in a single week its chemists reduced the level of martius yellow in their napthol yellow S from 10 to 0.25 percent; Kohnstamm's certified batch, however, contained only 0.05 percent. In part, Schoelkopf's success might have been a result of knowing it could be done, but it was also the result of having a staff of trained dye chemists. Where Kohnstamm had taken the best dyes it could purchase from Badische and developed methods for their purification, Schoelkopf took the best intermediates it could make or buy and fabricated its own dyes (47).

By the beginning of 1909, the food color question seemed to be heading rapidly towards a conclusion satisfactory to Wiley and Hesse. Two firms were competing to produce goods of hitherto unavailable quality, and one had begun to accept orders. The Board had accepted certificates for batches of all seven colors, so there were now standards for the colors, even if these were known only to Hesse, the Board, a few men in the Bureau's New York laboratory, and the certifiers themselves (48).

Dunlap and McCabe, as the majority of the Board, thought these standards should now be published for all to see. They instructed Hesse to prepare standards using the figures he had in mind when Kohnstamm submitted his colors. Hesse objected strongly. Early publication could only aid the foreign firms after two American businesses had done the work. He had had no figures in mind beforehand; his only idea had been to prod Kohnstamm and Schoelkopf to go as far as technically feasible. The two establishments deserved some reward for their efforts and cooperation when others had just scoffed. Allowing the firms to be the first on the market with certified colors, he argued, was the least the Board could do. While the reasons remain unclear, no standards were published at this time (49).

As the two companies approached the beginning of marketing, they found problems in the definitions of "mixture" and "batch" given in FID 77. Kohnstamm complained that traditionally food dyes had been sold to the trade mostly as blends diluted to some standard with a harmless diluent such as sugar or water, and that some customers, such as retail bakers, would only accept colors in paste or liquid form. Hesse agreed that since mixtures were to be allowed, these diluents should be permitted and recommended to the Board that they so amend the regulations. Schoelkopf argued that there was no logical reason to require each single preparation by a manufacturer to be a separate batch, especially as the equipment available often had a small capacity. Allowing uniform mixing of several such lots into a single batch seemed more reasonable. Hesse concurred, and passed this suggestion on to the Board. The Board accepted Hesse's recommendation but limited the size of such batches to five hundred pounds. It published these revised definitions in March 1909 as FID 106 (50).

Much to everyone's surprise, there was little initial market response to Kohnstamm's January announcement that it would accept orders for certified colors. E. G. Kohnstamm wondered if he would be left with only glory and financial loss for his efforts but pressed on; his company was publicly as well as financially committed and hoped that actual shipment of these colors and eventually the long promised Board announcement of a deadline mandating their use would spur sales (51).

Kohnstamm encountered the kinds of delays that often plague new systems and several times had to push back the initial shipping date. What it found most trying, perhaps because it was beyond its control, were the extensive delays in receiving certificates from the Board. The Board's procedure was cumbersome and slow. Kohnstamm sent certificates and samples to the Board in Washington where they were checked for form and completeness. These went next to the Bureau of Chemistry's New York laboratory where a single chemist, in addition to other duties, repeated all the tests the firm had done on the dye to see if it indeed matched the results and quality claimed. The chemist then sent his report to Hesse. If Hesse approved it, he informed the Board, who only then issued a certificate and assigned a lot number (52, 53). In exasperation, the company advertised in May that it would ship all orders on hand on June 1 "unless the US laboratory fails to check up our certifications . . . promptly, which has been the case up to now, but we have every reason to believe they will do more expeditious work on them in the future" (54).

This backlog was not cleared up until early July, so Kohnstamm did not publish its first price list of certified colors until August. All colors on this list were trade-named mixtures. Shoelkopf followed in September, advertising the seven colors as certified under their technical names (55).

Now that certified colors were in the marketplace, Hesse wrote Wiley in July, it seemed time for the Board to set a date after which the use of certified colors would be required. Wiley replied that he had already proposed to the Board that it set a date, perhaps October 1, after which it would proceed towards prosecution in all cases where other then certified colors were used. He had no idea, however, what action the Board, that is McCabe and Dunlap, would take. After consultation with both Kohnstamm and Schoelkopf, Hesse suggested January 1, 1910 as a more realistic date (56,57).

The Board flatly refused to accept this advice. When forced to a decision, the conflict between crusader Wiley and lawyer McCabe flared as it had over several other issues (58). With Dunlap casting the deciding vote, the Board ruled in November that, under the law, other dyes could not and should not be forbidden unless the Bureau of Chemistry (*i.e.* Hesse) could present evidence that the dyes in question were harmful to human health. Considering the number of dyes on the market, this was tantamount to rejecting mandatory certification. Still, the decision reached was inescapable to one concerned, as McCabe was, with the text of the law as enacted. The statute only expressly prohibited poisonous color in confectionary, and under the general adulteration clause, prohibited any added ingredient which would render a food harmful to health. This was as far as McCabe would go. The question of moral propriety was simply not relevant. It was the duty of the executive branch to enforce the law as enacted, and not to legislate (59).

Hesse doubted that a level of proof sufficient to convince McCabe existed for more than a few colors. The best that could be done would be to show that some specific coal-tar dyes were acutely toxic to test animals in relatively small doses. Many available colors had never been tested at all. Injurious character could be demonstrated for a few common impurities, especially arsenic. Even for the seven listed colors, all that could be said was that they had failed to cause harm in the face of attempts to demonstrate such harm in test animals. Justifying the quality control requirements of batch certification under McCabe's guidelines was even more unlikely. Hesse could argue only that, given a chance, the regulatory approach he had developed would provide an ample supply of safe colors for the industry and the public, while attempting to requlate colors by banning harmful ones would not (60).

McCabe and Dunlap apparently were convinced of the potential efficacy of the certification program, and were aware of the time and effort that had been expended in its development, but it remained a program without legal sanction. Through the winter, the two men searched for a way to resolve the conflict and salvage Hesse's work. Dunlap, a chemist himself, could well appreciate Hesse's arguments.

Both Kohnstamm and Schoelkopf were, as might be expected, incensed at the turn of events. The sale of certified colors thus far had been disappointing, disproving Kohnstamm's earlier contention that the certified colors would, within a few months, displace uncertified colors without any direct government pressure. W. H. Watkins, Schoelkopf's chief chemist, justifiably complained that the government consistently had led him and his company to believe that use of certified colors would be mandatory once the colors became available. In support of his allegation, he quoted from FID 76 and FID 77 and from several subsequent letters from Board chairman Wiley (61). What Watkins could not know was that Wiley was not in a position to deliver what he had promised. E. G. Kohnstamm, in similar disgust, wrote that his company would revive its uncertified line and would submit no further batches for certification "until such time as we can get the trade to believe that Certified Colors are the legal colors, or until some official action is taken to indicate to the general public that this is a fact" (62).

In the meantime, Dunlap and McCabe worked out what they believed was as good a solution as the law allowed. They would strongly and officially recommend the use of certified colors. Wiley refused to sign his name to this, but Secretary Wilson approved the scheme on April 7, 1910, and issued it as FID 117. The cleverness with which the FID was worded bears repeating:

"Certified dyes may be used in food without objection by the Department of Agriculture, providing the use of the dye in food does not conceal damage or inferiority . . .

Uncertified coal-tar dyes are likely to contain arsenic and other poisonous material, which, when used in food, may render such food injurious to health and, therefore, adulterated under the law.

In all cases where foods subject to the provisions of the Food and Drugs Act of June 30, 1906, are found colored with dyes which contain either arsenic or other poisonous or deleterious ingredient which may render such foods injurious to health, the cases will be reported to the Department of Justice and prosecutions had" (63).

To all but the most careful reader, this implied that the use of any color other than a certified one would be grounds for persecution. Only a very astute person, or a well trained lawyer, would realize that the government had retreated one step from mandating certification, and instead had merely reminded the trade that other colors were liable to contain harmful impurities and thus provide ground for prosecution. So, McCabe, Dunlap, and Wilson established the principle of a voluntary certification program for coal tar food colors coupled with strongly worded encouragement of its use. The government could not require use of certified colors, but it would attempt to make such use to industry's advantage.

While FID 117 may not have been the only reason, sales of

certified colors soon showed dramatic improvement. Other contributing factors included heavy promotion of certified colors by the two firms, and regulations issued by the state food commissioners of Illinois, Iowa, Idaho, and Nevada mandating the use of only certified colors in their states. In October, Kohnstamm reopened its production line. E. G. Kohnstamm wrote to Wiley of the great headway his firm had made in convincing food jobbers to market certified colors (64). The Board received enough requests from jobbers who wished to repack already certified colors under their own labels that it issued FID 129 in November, authorizing this practice and giving a procedure to be followed for recertification (65). By 1912 the demand for certified colors had become so great that Watkins requested that Schoelkopf be sent collect telegraph notification of acceptance or rejection of batches, just to save a day or two on the process, and by this time certification had become sufficiently routine that the time required had been reduced to an average of five days (66).

By early 1910, the bulk of Hesse's work had been completed. In June, he finally returned to the original work he had abandoned in 1907, the devising of a scheme for the detection, separation, and identification of coal-tar colors in foods. This work was also necessary as the government could not successfully prosecute a manufacturer for using a poisonous color in his products unless the dye's presence could be demonstrated. Over the next several months he completed this work with the assistance of a Bureau staff chemist. No report of this work was ever published (67).

Hesse also began to write an official report of his work for the Bureau. In it, he included a history of the project, the reasoning behind the regulations he developed, his entire literature search on the toxicology of the coal-tar colors (this was the longest section of the report), his analysis of commercial samples from 1907 of the seven permitted colors, the standards of purity for certification of each of these colors, and the proper methods for testing for these standards. He finished in late August of 1910 and sent the report to Wiley for review and comment. He made several minor revisions and additions in response to Wiley's comments, returned the manuscript to Wiley (68), and, his work completed, resigned his government post effective June 1, 1911 (69).

In January 1911, Wiley sent Hesse's manuscript on to Secretary Wilson with a recommendation that it be published. It appeared the following year as Bureau of Chemistry Bulletin #147, "Coal Tar Colors Used in Food Products." Hesse's report remained the standard work on the subject for many years; it was heavily cited by a British Parliamentary committee as late as 1954 (70).

On March 15, 1912, Wiley himself resigned, following Hesse out of government service. He had been worn down by years of fighting with Wilson and McCabe, and was weary of his frequent inability to establish enforcement policies, such as the mandatory certification of food colors, that he believed necessary for the protection of the nation's food supply (71).

Over the years that followed, an overwhelming majority of the primary dyes used in food came from certified lots. Both the number of batches and the total pounds certified grew steadily over the life of the Wiley Act. Thus the voluntary certification plan worked reasonably well. With the enactment of the Food, Drug, and Cosmetic Act of 1938, food color certification became mandatory. Although somewhat modified, in response to changed scientific standards, by the Color Additive Amendment of 1960, Hesse's procedures remain today the basis of American regulatory policy.

Food colors were one of the earliest products of scientific technology to be subjected to close government regulation, but over the years this type of control has been expanded to many other areas. As I have discussed at length elsewhere (72), the relative success of this program gives some cause for optimism. Over many years it has accomplished its goal: providing industry with an ample supply of products to meet technical and economic requirements while protecting the public from colors considered dangerous by ever changing contemporary scientific standards. While some critics have contended that the regulatory process does not and can not work, and others have documented cases where it has not worked, government by regulation is not doomed to failure. The case of food colors shows that is is possible to work with regulated industries, as Hesse and Wiley did with Kohnstamm and Schoelkopf, without abandoning the proper mission of the regulatory agency.

Acknowledgment

This paper is adapted from Chapter II of my Ph.D. dissertation at the University of Wisconsin-Madison, "Synthetic Food Colors in the United States: A History Under Regulation" (1982). I wish to thank my major professor, Aaron Ihde, and the other official readers of the dissertation, John Parascandola and Stanley Schultz, for their helpful suggestions for improving the original draft.

Literature Cited

1. Root, W.; de Rochemont, R. "Eating in America"; Wm. Morrow: New York, 1976; pp. 146-160, 213-246.
2. Beer, J. "The Emergence of the German Dye Industry": U. Illinois Press: Urbana, 1959.
3. Anderson, O. "The Health of a Nation"; U. Chicago Press: Chicago, 1958; pp. 120-196.
4. United States Department of Agriculture, Office of the Secretary, "Circular No. 21", October 16, 1906.

5. Wiley, H. "An Autobiography"; Bobbs-Merrill: Indianapolis 1930; p. 247.
6. Haynes, W. "The American Chemical Industry, A History"; 6 vols.; Van Nostrand: New York, 1948-1954; vol. 2, p. 61.
7. Oil, Paint, and Drug Reporter. February 9, 1917, 9.
8. Hesse to Wiley, 8/21/06. General Correspondence Files, Bureau of Chemistry, Record Group 97, National Archives. Washington D.C. All cited correspondence and other unpublished material is from this archive.
9. Hesse to Wiley, 11/15/06.
10. Wiley to Hesse, 12/1/06.
11. Hesse to Wiley, 2/6/07.
12. Anderson, reference 3, pp. 201-205.
13. Hesse to Wiley, 5/16/07.
14. Lieber, J. "The Use of Coal Tar Colors in Food Products"; privately printed: New York, 1904.
15. Weyl, T. "The Coal Tar Colors with Especial Reference to Their Injurious Qualities and the Restriction of Their Use"; Leffman, H. trans.; Blackiston: Philadelphia, 1892.
16. Fränkel, S. "Die Arzneimittel Synthese auf Grundlage der Bezeitungen Zwischen Chemische Aufbau und Wirkung", 2nd ed.; Springer: Berlin, 1906.
17. Parascandola, J. Pharmacy in History 1974, 16, 54-63
18. Green, A. "A Systematic Survey of the Organic Coloring Matters Founded on the German of Drs. G. Schultz and P. Julius"; MacMillan: London, 1904.
19. "Report of the work done by B. C. Hesse, June 10, 1907".
20. Hesse to Wiley, 6/15/07.
21. USDA, Off. of the Sec., Food Inspection Decision 76, 7/13/07.
22. USDA, Off. of the Sec., "Memorandum to Accompany Food Inspection Decision on Dyes, Chemicals and Preservatives"; n.d.
23. Hesse to Wiley, 7/27/07.
24. Bigelow to Hesse, 8/3/07.
25. Hesse to Dunlap, 8/8/07.
26. Hesse to Dunlap, 8/7/07.
27. Hesse to Dunlap, 8/16/07.
28. USDA, Off. of the Sec., FID 77, September 25, 1907.
29. Hesse to Wiley, 10/7/07.
30. Hesse to Wiley, 10/28/07.
31. USDA, Bur. Chem., "Bulletin 147", 1912, p. 1917.
32. Hesse to Shaw, 1/6/08.
33. Hesse to Wiley, 10/24/07, 10/28/07.
34. Lannen, T. Am. Food J. December 15, 1907, pp. 19-20.
35. Otten to Board of Food and Drug Inspection, 2/4/08.
36. Hesse to Wiley, 2/4/08.
37. Hesse to Wiley, 2/10/08.
38. Wiley to Price, 1/29/08.
39. Hesse to Bigelow, 11/21/07.
40. Hesse to Wiley, 2/8/08.
41. Hesse to Wiley, 2/28/08, 3/23/08.

42. Hesse to Wiley, 11/23/08, 1/4/09.
43. Advertisement, Am. Food J. February 15, 1909, p. 31.
44. Hesse to Wiley, 1/22/08.
45. Hesse to Wiley, 5/21/08, 5/25/08, 7/19/08.
46. Haynes, reference 6, vol 1, pp. 308-310, vol 6, pp. 292-293.
47. Hesse to Wiley, 5/26/08, 6/7/08, 10/8/08, 7/23/09.
48. Hesse to Wiley, 1/22/09.
49. Hesse to Wiley, 4/3/09, 5/27/09.
50. Hesse to Wiley, 3/11/09.
51. Hesse to Wiley, 2/9/09.
52. Kohnstamm to Wiley, 4/20/09, 5/6/09.
53. Wiley to Kohnstamm, 5/3/09.
54. Advertisement, Am. Food J. May 15, 1909, p 32.
55. Hesse to Wiley, 9/3/09, 9/23/09.
56. Hesse to Wiley 7/9/09, 8/25/09.
57. Wiley to Hesse, 7/28/09.
58. Anderson, reference 3, pp. 197-258.
59. Wiley to Hesse 11/27/09.
60. Hesse to Wiley, 12/1/09, 12/3/09.
61. Watkins to Board of Food and Drug Inspection, 1/7/10.
62. Kohnstamm to Board of Food and Drug Inspection, 1/8/10.
63. USDA, Off. of the Sec., FID 77, May 3, 1910.
64. Kohnstamm to Wiley, 10/22/10.
65. USDA, Off. of the Sec., FID 129, November 21, 1910.
66. Watkin to Board of Food and Drug Inspection, 3/3/12.
67. Hesse to Wiley 6/6/10, 8/26/10.
68. Hesse to Wiley, 8/26/10, 11/11/10.
69. Hesse to Wiley, 4/11/11.
70. Great Britain, Ministry of Foods, Food Standards Committee, "Report on Colouring Matter"; 1954.
71. Anderson, reference 3, pp. 252-258.
72. Hochheiser, S. Ph.D. Thesis, University of Wisconsin-Madison, 1982.

RECEIVED January 17, 1983

8

The Uric Acid Fetish: Chemistry and Deviant Dietetics

JAMES C. WHORTON
University of Washington, Department of Biomedical History, Seattle, WA 98105

Early twentieth century alarm over the presumed pathological effects of uric acid provoked considerable concern among both physicians and the public for controlling "uricacidemia" through diet and medication. Although finally discarded as merely a fetish, the uric acid scare is an impressive illustration of the impact of modern chemistry on popular, as well as medical thought.

"For many years 'diet' has taken the place of the 'weather' as a subject of conversation at the public dinner-table, and abstinence from certain foods has been raised to the level of a virtue by some people." (1) These remarks by an authority on gout and rheumatism were intended to describe the first two decades of the twentieth century. But they have as well a timelessness at once apparent to anyone who has endured contemporary dinner table conversation about vegetarian, organic, macrobiotic, megavitamin, or other exotic diet. Abstinence from certain foods and dedicated consumption of certain others, moreover, have been regularly justified throughout the past 150 years by arguments derived from chemistry. Around the turn of this century, particularly, as biochemical formulae became so abundantly available and the molecular philosophy of vital functioning rose to dominance in popular thought, dietary reformers exhibited a compulsion to rationalize their practices with hypotheses about reactions between the molecules of food and of the body.

To be sure, the formulation of those hypotheses was clumsy and subjective, guided less by scientific acumen than by determination to validate dietary practices already adopted on the basis of personal experience or philosophical inclination. Yet as scientifically naive and short-lived as food reform crusades have generally been, their leaders have too frequently attracted sizeable followings, including well-educated and otherwise critical minds, to be dismissed so flippantly as "the nuts among the berries." (2) One might even argue that in a sense the impact of

0097-6156/93/0228-0147 $06.00/0

chemistry on popular notions of nutrition has been more pervasive than in the more dramatic areas of medicine, agriculture, and warfare. The benefits of those latter activities have been gratefully, but passively received by the public; non-professionals have not been personally involved in their development, but have had those blessings conferred upon them as gifts from a chemical science beyond the layman's ken. Conversion to and practice of a dietetic heresy, however, requires an active effort to comprehend the biochemical mechanisms responsible for the practice's efficacy. The individual becomes involved in applying chemical theory to his own life, and misguided or not, that application is a significant consequence of the maturation of chemistry during the past century. A case study of an endeavor to utilize chemistry to purify popular diet can thus further illuminate the role of chemistry in the evolution of modern society.

Selecting the case to study is in itself an exercise in qualitative analysis. America's Progressive period was rich with dietetic diversity, the era's social reform optimism and enthusiasm for science pulling forth an extraordinary collection of proposals to save the world through corrected eating. Vegetarians, of course, had been drawing favorable (if invalid) conclusions from chemistry since the 1830s, but they took new heart from the late nineteenth century isolation of indole and other "ptomaines" as products of intestinal putrefaction. Autointoxication, or self-poisoning with ptomaines generated from a high protein diet, became the chief biochemical argument of early twentieth century vegetarianism. (3) It even influenced the thinking of meat-eating physicians sufficiently to force more than one professional leader to denounce the medical literature on autointoxication as "mad, maudlin, jumbled, mystic, undigested," and "sophomoric." (4)

A low protein diet, not necessarily meatless, was also strongly urged by Russell Chittenden and his followers. Professor of physiological chemistry at Yale and a pioneer in his discipline in the United States, Chittenden was led by personal experience and experiment to his conviction that 50 grams of protein per day was more healthful than the 120 grams then recommended by authorities. (5) But he explained his experimental findings with biochemical speculation about the effect of protein oxidation products on liver and kidney function, and one of his staunchest supporters, Harvard chemist Otto Folin, forwarded an elaborate scheme of protein metabolism to serve the same end. (6) Dietary restraint of another sort was advocated by Edward Hooker Dewey, a Pennsylvania physician who popularized a "No Breakfast Plan" of daily living, and employed fasting as a mode of hygiene and therapy as a result of his analysis of biochemical changes in the fasting body. (7) In addition, "apyrotrophers," the disciples of uncooked foods, accounted for the action of heat on food in terms of "proteids [being] coagulated,...the atomic arrangement of sugar...rendered incongenial,...and the oils... fused." (8) But of all the period's systems of deviant dietetics, none placed such heavy reliance on biochemistry or had so immediate and wide an effect on public, and medical, thought as the one despised by a critic as "The Uric Acid Fetish." (9)

Origins of the fetish

Few chemical compounds have so unenviable a history as uric acid. White, odorless, and tasteless, it is saved from being utterly nondescript only by its vulgar origins and pathological involvements. As a metabolic end-product, it is normally found in animal evacuations, in the urine of primates and the excrement of creatures such as birds, serpents, and slugs. It was discovered, in 1776, in urinary calculi, the kidney and bladder stones which have been the cause of so much suffering. And before the close of the eighteenth century, uric acid had also been identified in tophi, the deposits which torture gout victims' joints. In 1848, the London physician Alfred Garrod detected the substance in the blood of gout patients and suggested it to be the cause of the disease. (10) The final blows against the compound's character, however, were not struck until the close of the nineteenth century, when the acid came to be accused of an astonishingly long and varied list of medical crimes: "Of all words connected with disease," a commentator on *Fads and Feeding* wrote, "probably uric acid is more universally known than any other. One hears it continually spoken of as though it were the most familiar substance on this earth, and advertisements of some preparation which will seemingly expel this arch-fiend from the system meet one's gaze somewhere on the pages of every daily paper or periodical." (11)

The "High Priest of uric acid," as Clendening dubbed him, was another London physician, Alexander Haig. (12) Born in 1853, he was tormented by migraine throughout his youth and could find no relief in medicine. He finally turned, at age 29, to dietary treatment, first eliminating sugar, then tea, coffee, and tobacco. But all sacrifice came to nought until he swept his table clean of animal foods (except cheese and milk). Once the switch to lacto-vegetarian diet was made, he gradually overcame the headaches which had cost him one to two days of lost work each week during his student career at Oxford. The victory occasioned Haig's first scientific publication, an 1884 paper displaying already the cavalier handling of biochemistry which was to become his trademark. His migraine, he confidently theorized, had been due to impure blood, the impurity most likely being an alkaloid produced by intestinal putrefaction of excessive protein in his diet. His educated speculation took a new course, however, as Haig acquired clinical experience. He learned that several authorities believed migraine to be a consequence of a gouty constitution. He knew as well that the classic treatment for gout was a low protein diet such as the one which had removed his headaches, and that one of the common drugs in gout therapy--salicylic acid--was also utilized against headache. Closer examination uncovered several signs of gout in his personal and family history, so that by 1886 there could be little doubt in his mind that the migraine poison was none other than "our old friend uric acid." (13, 14, 15)

Development of the uric acid theory

That term "friend," intended to be facetious, was in fact prophetic. Haig was in the process of building a lifelong partnership with the substance, one to which he would devote extraordinary persistence and imagination in order to find a place for uric acid in every pathological situation. Born through personal experience, this dedication was to be sustained by his scientist's desire to simplify medical theory, and by an inadequate training in analytical and physiological chemistry which permitted him to invariably discover ways to make uric acid the simplifying factor. Every one of Haig's many papers was to have some connection to uric acid, as would his several books. His major work, *Uric Acid as a Factor in the Causation of Disease*, went through seven editions, each of which offered more than 900 pages of praise to the malevolence of the compound.

The friendship was too complex to allow discussion of all intimate details, but a general description of its evolution is easily given. Haig's suspicion that his headaches originated from uric acid was soon confirmed by analysis of his own, and other migraine patients' urine. Managing to find a direct correlation between the quantity of uric acid excreted and the occurrence of headache, Haig concluded that an excess of uric acid in the blood--uricacidemia--initiated headache, in which case the swallowing of a small dose of an acid to lower blood pH and reduce uric acid solubility should bring relief. Almost predictably the therapy worked, yet it did bring a surprise. The clearing of uric acid from the blood was succeeded by new symptoms of pains in the joints, symptoms suspiciously similar to those of gout. And when these pains were relieved by administration of alkali, Haig received his revelation, his sudden appreciation of the wonderful simplicity of what he began to call "the human test tube." (16)

In that test tube, he now saw, the poison uric acid could exist in solution or as a precipitate. Its condition depended on pH and temperature, just as in an ordinary test tube, and in either event, whether in the blood or tissues, it produced disease. That deposits in the tissues would cause irritation was obvious enough, but the mechanism by which dissolved uric acid worked its havoc required some logical analysis. Since his migraine pain was aggravated by stooping and alleviated by applying pressure to the arteries of the neck, Haig deduced the immediate cause of the headache had to be elevated blood pressure.

And since high uric acid levels were associated with migraine as well, there appeared to be a positive relation between acid in the blood and blood pressure. Finally, Haig had already observed that test tube solutions of urates could be chemically manipulated to form gelatinous colloids, and supposed it was "quite possible" a like phenomenon might occur in the blood. In one of his frequent blithe leaps from *in vitro* to *in vivo*, Haig postulated the condition of collaemia, "the presence of uric acid as a colloid in the blood stream and its action as an obstructor of the tiny capillaries." (17) By impeding the flow of blood, colloidal uric acid forced an increase in blood pressure throughout the body, and also, Haig reasoned, lowered the metabolic rate (or, to use his homely analogy, acted like "a wet blanket on a fire."). (18)

Application of the theory to pathology

This combination of depressed combustion and excessive blood pressure was a marvelously versatile construct which could be used to account for a remarkable assortment of clinical problems. Inhibited metabolism, for example, permitted sugar to pass through the body unoxidized, and was therefore the cause of diabetes. High blood pressure, on the other hand, was responsible for coma (after all, Haig observed, patients emerging from coma often complained of severe headache, and he had already shown headache to be produced by abnormal blood pressure). Comparable reasoning was employed to complete the list of collaemia diseases, a catalogue which ran through insomnia, asthma, menstrual dysfunction, neuritis, atherosclerosis, anemia, Bright's disease, Grave's disease, insanity, and hemorrhoids, to report but a few. Another which should not be left out of the recitation was the tell-tale collaemic face, a pitiful visage distorted by puffy skin and bulging eyes. Haig boasted he could guess a patient's blood pressure simply by the severity of his facial collaemia. (19)

Unfortunate though it was, collaemic face reflected only the ailments stemming from excess uric acid in the blood. Remaining were the problems caused by deposition of the compound in the tissues, the "local or precipitation group" of uric acid diseases. Gout headed this list, naturally, but all bodily tissues were subject to localized uric inflammation, so that problems ranging from gastritis and jaundice to eczema and flatulence could be laid to uric acid precipitates. The account was still not completed by the addition of these latter diseases, for it was evident to Haig that uric acid could be a predisposing factor as well as immediate cause. Microbic infection, the pathological entity with which most other physicians were preoccupied, might also be interpreted as a collaemic process. Germs, Haig submitted, thrive only in blood which is impure and sluggish, "while those falling into the brightly burning fire of a quick combustion with fine circulation and small accumulation of waste products are themselves burnt up and cannot produce disease." (20) Thus, he eventually came to maintain, "uric acid is the factor that controls the results of microbic invasion, and a life free from excess of uric acid is to a large extent a life immune from microbic injury." (21)

The final bit of wickedness in uric acid's nature was its ability to easily pass back and forth between blood and tissues, thereby subjecting its victims to alternating bouts with collaemic and precipitation ailments. Blood pH was one determinant of the substance's location: the more alkaline the blood, the more uric acid it would hold, and the more collaemic the victim would be. Highly alkaline foods, such as potatoes, he believed, would incite one set of diseases, acidic foods the other. Environmental temperature acted similarly, with heat pulling uric acid into the blood, cold pushing it out. With this neat scheme of pathological chemistry in hand, no clinical history and no epidemiological observation could escape Haig's analysis. Other doctors might have been baffled by the case of the Englishman who came down with an acute attack of gout after returning from a holiday in his homeland. Haig, however, discovered that while in England the patient had eaten heartily of foods containing the precursors of uric acid; that he had been bothered by the extreme heat on the Red Sea during his return, and then exposed to cool March winds in Bombay. What could be more certain than that the uric acid supplied by his English diet had been all drawn into his now pain-wracked joints by the chill of India? (22) Just as clear was the chemical foundation of the bit of folklore that "May is the month of suicides and murderers." Uric acid was largely restricted to the tissues during the winter because of cold weather and the English taste for oranges and other acid fruits during those months. But warm weather brought more alkaline diet, and thus a "spring cleaning" of the tissues. When uric acid rushed into the blood, exercising its collaemic effects on the brain and nerves, it brought on sudden melancholy or irritability. (23)

Even love was reduced to a uric acid equation. A young man's fancy turned the way it did in the spring, Haig intimated, because of his uric acid-induced blood pressure. Nature's release for that spring tension was sexual intercourse, for the exertion of the act decreased the blood pH, thereby precipitating uric acid and reducing blood pressure. Thus was a biochemical basis provided for the observation of statisticians that April, May, and June were the peak months for conceptions in Europe. That same theory also offered hope for suppressing the still disturbing perversion of masturbation. In Haig's view, the practice was an instinctive effort to relieve collaemic tension, and could never be controlled "with such feeble weapons as mental and moral suasion;" the miserable onanist, like virtually everyone else, was a prisoner of his circulation, and could be liberated only by purified diet. (24)

Onanism, melancholy, and gout, unfortunately, were but way-stations on the road to final catastrophe. The uric acid deposits in a person's tissues would gradually increase in quantity as the years passed and as his constitutional vigor waned, until at last, Haig predicted, "the long pent up store of urates breaks its dams and rushes into the circulation with an overwhelming flood." If not destroyed on the rocks of apoplexy, the helpless victim would be swept onward to heart failure, Bright's disease, or a like fate. (25)

One of the most terrible of those fates was cancer, an ailment Haig supposed to be a product of long term irritation by uric acid deposits. Cancer should therefore be responsive to dietetic therapy. In 1911-1912, Haig actually attempted to cure several cases of inoperable cancer with a diet (nuts, fruits, and biscuits) free of uric acid and its metabolic precursors. Although not one of the patients recovered, Haig refused to let his theory take the blame. The patients had either been too far gone when the diet began, or else lacked the intelligence and will to stay on the unfamiliar diet. "Hospital patients," he decided with some distaste, "are too ignorant or prejudiced to give this diet a fair trial in hospital, and they are too ignorant to provide what is required at home." A positive demonstration of his cancer regimen, he added, would have to be made "among the richer and better educated classes." (26)

"Uric-acid-free diet"

Haig nevertheless desired to make all classes better educated, to create a broad public awareness of the dangers of all foods containing uric acid or its purine precursors. His aim was conversion of humanity to the "uric-acid-free diet," though he realized that even that sweeping a purification would not immediately eliminate the poison. Uric acid is also produced by the body, through endogenous purine synthesis and catabolism, but just because it is endogenous, and thus a natural component of body chemistry, this quantity of uric acid seemed unlikely to be injurious. Haig anticipated, though, that endogenous uric acid production would ultimately disappear, as "the race evolves to a higher stage." (27) In the meantime, unnatural uric acid, that resulting from erroneously selected food, had to be dealt with. The first step, of course, was identification of offensive foods, a process which conveniently incriminated the animal foods Haig had already abandoned on the basis of personal experience. By the mid-1890s, he also recognized that chemical data branded as dangerous several vegetable foods he had continued to eat: peas, beans, asparagus, mushrooms, and whole grain cereals also had to be rejected because of their purine content. Haig was left, then, with milk, cheese, some vegetables, fruits, nuts, and--a unique position for a food reformer--white bread. Additional blandness was supplied by the prohibition of coffee, tea, and cocoa on the grounds that they contained methyl xanthines (it was later found that caffeine and similar compounds are not metabolized into uric acid). Any rejoicing that at least alcoholic beverages were free of uric acid-producing substances was immediately squelched by Haig's assurance that his diet eliminated any need for stimulation and thus destroyed the craving for strong drink. (28)

Uric acid and athletics

Haig clearly appreciated few people would be driven to such austere fare by gustatory impulses, for while he promised readers they could learn to enjoy the uric-acid-free diet, he directed far more attention to the pleasures of uric acid free life away from the table. A disease-free, physically joyful existence, a doubled life-expectancy, a pain-free death--all these were the usual rewards of the diet. But Englishman that he was (and having been a rower at college), Haig showed the greatest excitement over the sporting benefits of his dietary program. *Diet and Food* was his contribution to an already torrid debate over the connections between nutrition and athletic performance. The traditional training diet, loaded with meat, was being called into question by an army of muscular vegetarians who were regularly thrashing flesh eating rivals at cycling, walking, running, tennis, and other competitions. The issue was only complicated further by Haig's book, which went through six editions in the years from 1898 to 1906, and predictably revealed that even athletic success was an expression of freedom from uric acid. *Diet and Food*'s full theory of strength and endurance, however, must have startled even the most jaded Haig-watchers. The premise on which the entire book was constructed was the idea that the energy for muscular motion was provided by the oxidation of protein to urea. That notion, originally proposed by Liebig half a century before, had long since been refuted and was no longer regarded seriously--except by Haig. He seems to have been unaware of his loneliness on this position, however, presenting the theory as if it were generally accepted, and even applying his inimitable analytical skills to the discovery of a direct correlation between quantity of exercise and excreted urea. If the energy for exertion came from protein, Haig went on to hypothesize, then maximum strength and endurance demanded a free flow of protein-rich blood to the muscles. Vessels clogged with colloidal uric acid, of course, would neither be able to provide a full complement of protein molecules to the tissues, nor to carry off the waste of protein oxidation. The more uric acid food an athlete consumed, therefore, the more physiological "friction" he would suffer, and the lower would be his performance in contests of endurance. (29) That was undoubtedly why flesh eaters usually succumbed to vegetarian challengers, but vegetarians, he continued, had no reason to celebrate. Their vessels too were polluted, for beans, asparagus, and mushrooms yielded uric acid. Ordinary vegetarianism granted a relative advantage, but uric-acid-free vegetarianism was the only means to absolute superiority. (30) This argument was reinforced by the record of the most successful pedestrian of the day, Karl Mann. A long distance walker who had switched to standard vegetarianism in 1894, Mann converted to Haig's refined version in 1898. He won an important 70 mile race within the year, but it was the 1902 Dresden to Berlin triumph in world record time which elevated Mann into the international spotlight. (31) Haig hurried to Berlin, personally examined Mann just a few days after the race, and was pleased (but not surprised) to find him free of the cardiac hypertrophy believed to be epidemic among carnivorous competitors. "Athlete's heart," he concluded, was still another uric acid-produced ailment. (32)

Uric acid and society

Haig never used the term, but he also believed in a condition of "uric-acid-free heart." By this he would have intended heart in the complete meaning of the word: the possession of strength, courage, hope, and mental energy. Haig's vision of total health, the goal of his diet, was of a state in which "exercise of mind and body is a pleasure, the struggle for existence a glory, nothing is too good to happen, the impossible is within reach, and misfortunes slide like water off a duck's back." In the end, Haig, like other health reformers, found moral force in his system, and came to see it as a method for the perfecting of individuals and, consequently, the nation. His writings, in fact, carry a feeling of urgency for its application to social challenges such as physical and moral degeneracy, the falling birthrate in the upper classes, and the decline of Britain's imperial standing. Ultimately, he hints, uric acid could be responsible for nothing less than the disintegration of English civilization. (33)

Two biochemical mechanisms for cultural decay were identified by Haig. First, collaemia, by inhibiting circulation to the brain, could prevent clarity of thought, sap mental energy and endurance, and enfeeble will power. The social results of the subjection of an entire citizenry to cerebral starvation were highly unsettling to consider. No less frightening were the effects of uric acid's second mode of action, its generation of an appetite for stimulants (though a theoretical mechanism for this effect was extracted from biochemistry only after the most ruthless torture). So arcane as to defy explanation except in its naked essentials, this argument took its start from the metabolic depression produced by uric acid. Physiological stimulation, it continued, requires a clearing of the compound from the bloodstream. Experiment indicates that meat, tea, coffee, alcohol, and opium are all uric acid precipitants, hence are stimulants. The stimulation they cause, however, is only temporary, as the precipitated acid soon redissolves, in greater quantity than before, and makes the victim more depressed than ever and with a renewed, stronger craving for stimulation. The more meat, tea, and coffee one takes, the more he wants, until finally meat alone can no longer satisfy the need. "And when meat begins to fail," Haig sadly concluded, "alcohol is added; when alcohol begins to fail morphine or cocaine are called in and so on down the road to ruin." (34)

Uric-acid-free diet would remove any craving for alcohol, tobacco, or other stimulants, but making this plan of character rescue attractive to people already enslaved to stimulation, already debauched by uric acid, was a formidable challenge to even Haig's ingenuity and optimism. He sometimes envisioned a future "which will be...truer, nobler and better, as man slowly realizes how much of his sordid past has had its origin in unnatural food." (35) His more common mood, though, was gloom. Scattered through his writings are calls for "the nations which hope to survive" to open their eyes and see that uric acid "bids fair to menace our very existence." But would they listen? "No! I fear not! for history shows that things of this kind have been discovered and forgotten, rediscovered and reforgotten, and no doubt the process will be repeated yet many times; still I do think that possibly the representatives of *homo sapiens* (not of this race, for it will be mostly wiped out), in the 30th or 40th century may be a little more unanimous than they are today in believing that their natural food is after all that which is also best for them." (36)

It was Haig's son, however, who gave the most complete expression to anxiety about the future of the nation. Kenneth Haig, like his father, was plagued by migraine from an early age, and then, when a sixteen year old at Rugby, began to suffer from fainting spells. Adoption of the uric-acid-free diet brought a steady release from his problems, though, as well as a growth of endurance which astonished even Haig *pere*. "At Oxford he won several college rowing prizes," father bragged, "and would, I believe, have been in one of the college boats, but that it was feared that his diet would demoralize...the rest of the crew." (37)

Kenneth's filial debt was unusually large, and he found deep satisfaction in partially repaying it with a 1913 volume entitled *Health Through Diet*. The title page also acknowledged his father's advice and assistance, so one might presume that Kenneth's comments reflected the concerns and opinions of Alexander. If so, both were greatly worried by the country's declining birthrate, and suspected uric acid as the cause. The younger Haig postulated that one clinical manifestation of uric acid precipitation must be an irritation of the vaginal wall, a "catarrh of the vagina" which dampened sensitivity and led to sexual apathy. As the degree of indifference should be proportional to the amount of meat and other uric acid foods consumed, fecundity could be expected to show an inverse relation with rising social position--exactly the disturbing pattern which was being observed. Uric-acid-free diet, however, should enhance fertility by restoring natural function, a theoretical proposal confirmed by the woman who had been pronounced sterile by her physician, but then conceived and bore a healthy child two years after going on the Haig diet. Other women on the diet reported decreased menstrual flow and easier periods, prompting Haig to wonder if ovulation "need...be accompanied by hemorrhage, or is the latter only a pathological symptom?" After also considering that elimination of uric acid would relieve morning sickness, allow less painful labor and safer birth, and promote lactation, he felt compelled to call attention to the "national importance" of the "influence of the Uric-Acid-Free Diet on gynaecology and midwifery." His preface went further, proclaiming "that the rise and fall of nations is determined by the circulation." Nations, like armies, march on their stomachs, and "in the last resort their commisariat is their success or ruin." (38)

Uric acid and the medical profession

The investing of uric acid with cosmic import stretched plausibility too thin for all but the most credulous. The simpler idea that it caused a number of purely physical ills, though, was able to win respectability in some medical circles and hold it for a good while, and it was at least as popular in the United States as in England. To be sure, the Brahmins of the profession were scornful from the start. By the early 1900s, in fact, articles *contra* Haig made up a distinct genre of medical literature, with remarkably consistent content. The comments of the editor of the *Journal of the American Medical Association* are representative: "Using methods that are known to be unreliable, he secured data that cannot be corroborated, and with these as a basis followed out a marvelous train of logic to the ultimate conclusion that practically all disease is due to uric acid....No advance in physiologic chemistry and pathology, no amount of refutation of his claims, seems to have interested him or swerved him in the least,...he ignores everything but his own cherished beliefs, and calmly follows them as they lead." (39)

But as frustrating as Haig's inattention to these frequent scoldings could be, uric acid critics were more bothered by the naivete of rank and file physicians, so many of whom, it was charged, had allowed themselves to be conditioned to reflexly diagnose "uric acid diathesis" if any uric acid were found in the patient's urine. That the compound had been able to acquire what one commentator described as a "fixed hold" on the medical profession should not, however, have caused much surprise. (40)

Even the best educated medical scientists of the day had still only a nebulous view of metabolism and blood chemistry, yet all the while exuded confidence that both areas would be demonstrated to have important relations to disease. Standard pathology texts included discussion of the gouty diathesis, a loosely defined aggregation of symptoms commonly found among patients who never suffered an acute attack of gout, but had a family history of the disease. Uric acid was thus surrounded by just enough murkiness to give Haig's ideas an appearance of substance to the average medical practitioner with a tenuous hold on biochemistry. In addition, Haig's articles were published in prestigious journals--_Lancet_, _Practitioner_, _British Medical Journal_, _Medical Record_, and _Journal of the American Medical Association_. His books were impressively thick volumes, written in technical language and replete with detailed graphs relating uric acid excretion to the ingestion of various foods and drugs, to exercise, menstruation, time of year, and even to the effects of a Turkish bath. His _magnum opus_ was revised six times in order to incorporate the results of his continuing researches. By supplying a simple, unifying theory, Haig satisfied the needs of harried physicians who encountered on a regular basis patients with indeterminate symptoms and no obvious pathology. Giving the devil his due, an American medical editor marvelled that, "no promoter of a commercial enterprise has ever been more skillful in enunciating, and pushing to a conclusion, theories as to the existence of a precious find in a mine than has Dr. Haig." (41)

Thus there is no cause for wonder that in spite of authoritative refutations, Haig attracted a sizeable band of supporters. As early as 1901, a leading American journal contemptuously identified uricacidemia as _the_ disease of the new century: "Now that 'malaria' is coming to be known among the backward members of the profession as a specific disease caused by a parasite..., a need is felt for some new catchword with which to mystify ourselves and the public. Judging from the glibness with which the 'uric acid diathesis' is now talked of by both doctors and patients, this is apparently the coming disease. It explains numberless strange symptoms, gratifying the sufferer and having withal a comforting, scientific sound, which promises well for its usefulness and permanence. We know of physicians who are constantly making the diagnosis with pride, and who have not even dreamed that the matter needs any verification in the chemical laboratory. For these men it may be well to say that not by any means as much is known of the role of uric acid in the animal economy as might be supposed from the extravagant writings of Dr. Haig." (42)

Woods Hutchinson declined to apologize for mounting yet another attack on the uric acid theory because, he warned, "if we do not take it, it will take us." (43) The fact that other medical celebrities, including Lafayette Mendel, (44) Lewellys Barker, (45) and J. J. R. Macleod, (46) took the time to second those sentiments suggests that belief in the theory was indeed common, though we have the testimony of the believers themselves to finalize the point. Journal articles asserting the hazards of uric acid made free use of adjectives such as "lethal," "insidious," and "hydra-headed." One doctor reported swallowing a few drops of acid and immediately feeling a flood of uric acid gushing into his big toe. (47) A Massachussetts physician advised heroic therapeutic measures to save a two month old child from uric acid poisoning, (48) while a New York doctor pronounced that "uric acid was the Devil himself." (49) A final writer commented on a cartoon "in which a prominent society woman is represented as saying emphatically to one of her lady friends: '*I do believe that I am just lousy with uric acid!*' Probably she was. Many of us are." (50)

Difficulties of the uric acid diet

As the last example suggests, the publicity given the uric acid theory also set off an epidemic of self-diagnosis by the public. "Indeed," a Chicago physician complained, "among the laity it is the almost universal belief that a joint pain, myalgia, neuritis, neuralgia, etc. are due to uric acid....How often does a patient say: 'My doctor says I have uric acid.'" (51) By 1902, advertisements for Quaker Oats were referring to Haig to assure consumers the cereal would give "Home-Made Health." (52)

Whether or not Haig actually ate Quaker Oats himself, he did agree that diet was the only method of defeating uric acid. He also recognized most people would have difficulty converting to a uric-acid-free diet, due to habit, because past indiscretions had established a taste for continued stimulation, and because his regimen was commonly confused with ordinary vegetarianism. There was no end to Haig's irritation, in fact, over many people's inability to distinguish between the garden variety vegetarianism (which in its "ignorance" produced only "unfortunate results") and his selective "physiological and purin-free diet." But even though he complained "vegetarianism has been a great thorn in my side," standard vegetarians reacted favorably to his work, interpreting it as primarily a condemnation of meat eating. (53)

To make the transition from regular to refined diet easier, Haig developed a plan of gradual withdrawal from uric acid. Son Kenneth suggested a schedule of removing uric acid from breakfast only for three weeks, then from lunch as well for the next three weeks, and so on. This plan included as one stage the withdrawal of all tea from afternoon tea. Followers also published at least two uric-acid-free cookbooks, (54, 55) while Haig opened a Sanatorium (Apsley House) at Slough, about half an hour from London by train. A gift from a grateful patient, it was "a fine Christopher Wren house, standing in lovely old-world grounds," outfitted with garden and greenhouse, and adjacent to scenic countryside and three golf courses. (56) Apsley House offered an ideally congenial environment for the return to natural diet, but it was accessible to relatively few, and fewer still took advantage of it, or of the diet it offered. By Haig's own estimate, only a few hundred had "dared" to adopt the diet, and his detractors implied the number was even lower. (57) The uric-acid-free diet was, in the common view, "a very joyless one, as well as being socially a nuisance." (58) That valuation was, if anything, corroborated by those who actually followed the dietary plan. A correspondent of the British Medical Journal who identified himself only as "A Quondam Gourmet," confessed that, "I am a convert [to Haig's diet] in spite of myself." "I still," he elaborated, "hanker after the fleshpots. Caneton Rouennaise, with a bottle of Chambertin, still appears more attractive than Apsley Duck with salutaris, and English roast beef more savoury than mock beef rissoles. Though my memory dwells with pleasure on many a past gastronomic treat, yet the improvement in my health and the increase in my power of endurance are such that nothing would induce me to revert to my former dietetic habits; and I know that mine is far from being a solitary experience." (59) Such resigned followers of Haig were variously presented as "cranks," "martyrs," and "shrivelled, juiceless, prematurely aged" creatures; but they were never described as numerous. (60, 61)

Uric acid therapeutics

So as frightened as many people were of Haig's disease, the majority clearly regarded Haig's cure as still more discomfiting. There was literally a ready solution to the dilemma, however. The therapy of gout had long included the use of drugs supposed to be capable of dissolving uric acid and washing it from the body. Lithium compounds enjoyed an especially high reputation, having been recommended by Sir Alfred Garrod himself after he observed lithium carbonate to have a powerful solvent effect on uric acid *in vitro*. (62) But even though it had been determined by the beginning of the twentieth century that lithium's solvent action did not occur inside the body, doctors continued to prescribe the carbonate, and other lithium salts, to treat gout and gouty diathesis. Thus as uricacidemia established itself as a prevailing syndrome, lithium ascended to the position of panacea. Medicine manufacturers stampeded into the new market, vending every imaginable synthetic preparation, in addition to a line of natural, lithium-containing mineral waters. One had the cheek to name his product "Garrod Spa," and the proprietors of another --Thialion--actually published their own journal, *The Uric Acid Monthly*, and distributed it gratuitously to all physicians in the United States and England. *El Acido Urico* was added later for the benefit of Spanish and Latin American doctors. (63) Although other uric acid solvents were less aggressively promoted, they were promoted nonetheless, leading some physicians to charge that drug industry literature, embellished by detail men, was more important than Haig's writings for creating professional and public fear of uric acid. (64) The irony of that development was that lithium compounds had no effect on uric acid, and even if they had, few preparations contained enough lithium to matter. Thialion's formula included less than two per cent lithium citrate, and the highly touted Buffalo Lithia Water, it was eventually determined, held only one-fifth the concentration of lithium occurring in the Potomac River; a patient would have had to drink at least 150,000 gallons of the mineral water daily to get a therapeutic dose! (65)

Even before the impotency of lithium preparations became common knowledge, though, there were physicians who voiced doubts, sometimes tongue-in-cheek. A 1895 parody--"Twinkle, Twinkle, Garrod Spa"--for instance, begged the lithium water to

"Allay my fears, relieve my pains
By clearing crystals from my veins."

"Thou dost," the poet eventually thanked Garrod Spa,

".....dispel all carking care,
Bring back my youth so debonair,
Make me happy, calm and placid
By chasing out the uric acid." (66)

As such attacks continued, uric acid's standing with the profession as a whole was steadily lowered. By 1915, physicians could joke that, "There was as much to be said for uric acid as there is for intestinal toxemia [the invalid theory of ptomaine poisoning]: it cured for a time its thousands, but the theory built about it caught cold and died." (67) Already the year before, an American journal had concluded its brusque review of Kenneth Haig's Health Through Diet with the prediction that "the uric acid fad has had its day, and...the present volume will not resuscitate it." (68)

The younger Haig's volume was indeed the theory's last gasp. Alexander Haig also gave up the struggle after 1914, spending his final decade in quiet retirement. It would be wrong, however, to conclude that he had been cashiered from the profession. Even in his most passionate anti-uric acid days he commanded personal respect, and even affection, from professional colleagues. On friendly terms with some of the brightest lights in British medicine, Haig was regarded as honorable, and as generally competent when not astride his uric acid hobby horse. He was seen as quixotic, rather than quackish, and even the hard condemnations of his work often showed a soft side, whether expressions of gratitude for raising questions and stimulating research, or praise for his integrity and perseverance. His obituary notices sounded the same tone, one describing him as "a man genial in nature ...[who] always seemed to maintain an even mental equilibrium under adverse criticism." (69) The best summary of his life, however, was contained in one of Haig's own incidental remarks. The rhetorical question, "Is life worth living?," he wrote, is not adequately answered by the French response that it depends on the liver; actually, he announced with unintended prescience, "that depends upon uric acid." (70)

Acknowledgments

This article is adapted and revised from material in James C. Whorton, Crusaders for Fitness: The History of American Health Reformers.

Literature Cited

1. Wilde, Percy, "The Physiology of Gout, Rheumatism and Arthritis"; Wood: New York, 1922; p 2.
2. Deutsch, Ronald, "The Nuts Among the Berries"; Ballantine: New York, 1961.
3. Kellogg, John, "Autointoxication or Intestinal Toxemia"; Modern Medicine: Battle Creek, Mich., 1919.
4. Clendening, Logan. Interstate Med. J. 1915, 22, 1192-3.
5. Chittenden, Russell, "Physiological Economy in Nutrition"; Stokes: New York, 1904.
6. Folin, Otto. Am. J. Physiol. 1905, 13, 117-38.
7. Dewey, E. H. "The No-Breakfast Plan and the Fasting-Cure"; Dewey: Meadville, Pa., 1900.
8. Drews, George. "Unfired Food and Tropho-Therapy"; Drews: Chicago, 1912.
9. Miles, E.; Collings, C. H. "The Uric Acid Fetish"; Miles: London, 1915.
10. Kelley, W.; Weiner, I., Ed.; "Uric Acid"; Springer-Verlag: Berlin, 1978; p 1-11.
11. Read, Stanford. "Fads and Feeding"; Dutton: New York, 1909; p 94.
12. Clendening, Logan. Interstate Med. J. 1915, 22, 1199.
13. Haig, A. "Uric Acid as a Factor in the Causation of Disease"; Blakiston: Philadelphia, 1903. p 1-4.
14. Haig, A. Practitioner. 1884, 33, 113-8.
15. Haig, A. Practitioner. 1886, 36, 179-80.
16. Haig, A. "Uric Acid as a Factor in the Causation of Disease"; Blakiston: Philadelphia, 1903. p 176.
17. ibid. p 174.
18. Haig, A. "Uric Acid. An Epitome of the Subject"; Churchill: London, 1906. p 37, 54.
19. ibid. p 49-60.
20. ibid. p 105.
21. Haig, A. Med. Rec. 1910, 78, 391-4.
22. Haig, A. Brit. Med. J. 1895, 2, 1605.
23. Haig, A. Med. Rec. 1905, 68, 332-5.

24. Haig, A. "Uric Acid as a Factor in the Causation of Disease"; Blakiston: Philadelphia, 1903. p 154-6.
25. ibid. p 111-2.
26. Haig, A. Brit. Med. J. 1912, 2, 82.
27. Haig, A. "Uric Acid as a Factor in the Causation of Disease"; Blakiston: Philadelphia, 1903. p 838.
28. Haig, A. "Diet and Food"; Churchill: London, 1902. p 47.
29. ibid. p 2-6.
30. Haig, A. "Uric Acid as a Factor in the Causation of Disease"; Blakiston: Philadelphia, 1903. p 830-5.
31. Caspari, Wilhelm. Pfluger's Arch. Physiol. 1905, 109, 569.
32. Haig, A. Med. Rec. 1906, 69, 817-9.
33. Haig, A. "Uric Acid as a Factor in the Causation of Disease"; Blakiston: Philadelphia, 1903. p 321.
34. ibid. p 874.
35. ibid. p 136-7.
36. Haig, A. "Uric Acid. An Epitome of the Subject"; Churchill: London, 1906. p 99.
37. Haig, A. "Diet and Food"; Churchill: London, 1902. p v.
38. Haig, K. "Health Through Diet"; Methuen: London, 1913. p 117-23.
39. Ed. JAMA. 1904, 43, 550.
40. Goodhart, J. Practitioner. 1906, 76, 9.
41. Ed. Ther. Gaz. 1906, 30, 181.
42. Ed. Pediatrics. 1901, 11, 261.
43. Hutchinson, Woods. Lancet. 1903, 1, 288.
44. Mendel, Lafayette. JAMA. 1906, 46, 843-6, 944-7.
45. Barker, Lewellys. "Truth and Poetry Concerning Uric Acid"; American Medical Association: Chicago, 1905.
46. Macleod, J. J. R. Cleveland Med. J. 1905, 4, 465-73.
47. Uric Acid M. 1902, 2, 404.
48. Uric Acid M. 1906, 6, 21.
49. Uric Acid M. 1911, 11, 16.
50. Berkart, J. B. Brit. Med. J. 1917, 2, 208.
51. Billings, Frank. Am. Med. 1901, 2, 565.
52. Dietetic Hygienic Gaz. 1902, 18, xix.
53. Haig, A. Lancet. 1908, 2, 1632.
54. Haig, G. "Some Recipes for the Uric-Acid-Free diet"; Haig: London, 1913.
55. Webster, Mrs. J; Jessop, Mrs. F. W. "The Apsley Cookery Book"; Churchill: London, 1905.
56. Haig, A. "Uric Acid. An Epitome of the Subject"; Churchill: London, 1906. p 146-7.
57. Haig, A. "Uric Acid as a Factor in the Causation of Disease"; Blakiston: Philadelphia, 1903. p 837.
58. Ed. Brit. Med. J. 1908, 2, 1781.
59. Eve, F. C. Practitioner. 1905, 75, 823.
60. Allbut, T. C. Practitioner. 1903, 71, 1-5.
61. Goodhart, J. Practitioner. 1906, 76, 11.
62. Ed. Ther. Gaz. 1903, 27, 242-3.

63. Ed. JAMA. 1906, 46, 1381-2.
64. Ed. JAMA. 1906, 46, 1113.
65. American Medical Association. "The Propaganda for Reform in Proprietary Medicines"; American Medical Association: Chicago, 1916. p 205, 468.
66. Ed. Med. Mirror. 1895, 6, 582.
67. Clendening, Logan. Interstate Medical J. 1915, 22, 1199.
68. Ed. Ther. Gaz. 1914, 38, 451.
69. Ed. Lancet. 1924, 1, 826.
70. Haig, A. "Uric Acid as a Factor in the Causation of Disease"; Blakiston: Philadelphia, 1903. p 322.

RECEIVED December 1, 1982

9

Chemical Warfare Research During World War I

A Model of Cooperative Research

DANIEL P. JONES

University of Illinois at the Medical Center, Humanistic Studies Program, Chicago, IL 60612

Project-research, a method of organizing research by stipulation of projects and allocation of these to individuals or teams of scientists in separate laboratories, was developed in the United States during World War I in research on chemical warfare. This research was initially conducted largely by academic chemists as volunteers and later by them in the Research Division of the Chemical Warfare Service of the U. S. Army. Many of the leading American chemists in the 1920s shared the common experience of research on chemical warfare. The model of project-research was tried by the leaders of the division of chemistry and chemical technology of the National Research Council in order to allocate specific research problems and foster cooperative research after the war.

The dramatic transformation that the sciences have made on modern warfare became apparent during World War I. Just after the war Robert M. Yerkes edited a volume entitled The New World of Science: Its Development During the War, in which several scientists described well the transformation they had witnessed and the roles of various sciences in bringing about this change. Moreover, a few of the authors recognized that the war had required the mobilization of the scientific resources of the nation to an extent not previously encountered and that the results of many teams of scientists working on coordinated programs of governmentally sponsored research had been impressive. For example, George Ellery Hale, James R. Angell, and Robert A. Millikan cited the productivity of such teams as evidence for the power of cooperation in research, and they called for a similarly coordinated approach to scientific problems in peacetime (1, 2).

In the decade following the war, the leaders of the National Research Council (NRC) assumed major responsibility for organizing

0097–6156/83/0228–0165 $06.25/0

scientific research in the United States. Through Hale's initiative the NRC was reorganized as a permanent part of the National Academy of Sciences, dedicated to the stimulation of pure and applied research and the promotion of national and international cooperation in research (3). In order to foster cooperation the NRC was "to survey the larger possibilities of science, to formulate comprehensive projects of research, and to develop effective means of utilizing the scientific and technical resources of the country for dealing with these projects" (4). This approach, "project-research", which had proved to be effective in achieving cooperation among scientists in wartime was carried over into the postwar period by the NRC in several of its divisions. For example, the chairmen of the division of chemistry and chemical technology attempted to guide chemical research during the decade from 1919 to 1929 by the assignment of projects to volunteers. The model that served well was the organization of wartime research on chemical warfare agents, the largest of the governmentally sponsored research programs of the war, initially under the Bureau of Mines and later as the Research Division of the Chemical Warfare Service (5). Over one-tenth of the chemists of the United States were directly involved in the chemical warfare research effort in World War I (6). Research on chemical warfare was singled out as a model of project-research after the war because of its size, the success with which it guided work on interrelated projects, and also because many of the chemists, pharmacologists, and physiologists formerly in the Chemical Warfare Service held positions of leadership in the scientific community in the ensuing decade. Although it provided inspiration for project-research in several fields of science, the model of chemical warfare research was especially strong on American chemists because they felt a special kinship with the Chemical Warfare Service of the U.S. Army. This kinship was an important factor in the failure in 1926 of the United States Senate to ratify the Geneva Protocol prohibiting chemical warfare (7).

This chapter will examine the nature of project-research as it developed in the organization of chemical warfare research during World War I and will suggest that this model may have played a significant role in the attempts at increased organization of chemical research in the United States after the war, especially in the division of chemistry and chemical technology of the National Research Council.

Project-Research

Project-research, as it evolved during the war and through the 1920s had several characteristics common to what is known as the "problem approach" already in existence at that time. The United States government had pioneered in the problem approach as early as 1880 in the Department of Agriculture, and for years several bureau chiefs had assigned experts in various disciplines to work together on specific problems with little regard to traditional boundaries of scientific disciplines (8). The problem approach had been evident also in the work of the Rockefeller Institute and the Carnegie Institution since their establishment in the first decade of the century. However, during World War I in attempting to expand this approach to a sizeable portion of the scientists of the nation, new difficulties were encountered. First, the number of investigators to be organized was quite large, and second, most of them were not employed by the federal government, but had to be recruited as volunteers to work for little or no remuneration. Wartime project-research resembled the earlier problem approach in that it proceeded by the stipulation of research problems worthy of attention, allocation of these to scientists, collection of their results, and suggestion of further projects. Project-research differed from the problem approach in that the scientists were generally unpaid volunteers and were selected from a large population that were working in various colleges and universities at considerable distances apart. These distinctions became clear after the war when, despite the nearly general agreement among scientists on the desirability of cooperation, there prevailed a fear that such schemes might interfere with the traditional independence of scientists. Project-research was considered acceptable because the autonomies of the investigators were maintained and their projects were limited to specific problems of small size.

Henry Prentiss Armsby, chairman of the agriculture section of the American Association for the Advancement of Science, urged the members of the AAAS in 1919 to avoid being stampeded by the success of wartime scientific research into an undue exaltation of the virtues of scientific cooperation and organization. He explained that

> The men who worked together almost night and day to devise efficient gas masks or means of submarine detection or methods of sound ranging were not workmen under the orders of a superior, but free associations of scientists with training in common or related fields of research and under the inspiration of a common patriotism. Precisely this is what is needed to achieve the victories of peace. Effective cooperation cannot be imposed from above by administrative authority but can only come by free democratic action of investigators themselves. (9).

The extension into peacetime of project-research required a commonly held ideal similar to the "inspiration of a common patriotism." Armsby believed this could be found in the program of the newly reorganized NRC, in that its plan for cooperative research was based upon the voluntary initiative of investigators, united by their common interest in solving problems essential for the advancement of science in the United States. Because the leaders of the NRC were convinced that this attitude prevailed among American scientists, they were quick to assure them that the NRC's program did not restrict the freedom of scientists. Representative of many such statements from 1919 to 1921 is that of Vernon Kellogg, permanent secretary of the NRC:

> I know of no one in the National Research Council ... who dreams of suggesting the advisability of organizing or in any way interfering with, the individualistic work of scientific genius. What is suggested as advisable, because it was proved to be possible and highly effective in our wartime efforts, is to arrange for planned, concerted attack on large scientific problems as require numerous cooperating workers and laboratories representing, often not alone one special field or even one major field or realm of science, but several such fields (10).

Project-research was essential to the plans of the NRC not only to ensure scientific freedom but also for a financial reason. Having no substantial funds to support research itself, the NRC was dependent upon scientists volunteering their efforts. In both respects the model of organization used by the research program on chemical warfare during the war was found to be a very useful one by the leaders of the NRC as they attempted to organize American science in the 1920s.

Voluntary Nature of Research on Chemical Warfare

The mobilization of chemists for research on chemical warfare was accomplished almost entirely through their own voluntary efforts and not at the instigation of the military. In fact, military leaders in the United States initially had ignored the need for any chemical research in this area, although many chemists at the time recognized this need, as did the director of the Bureau of Mines.

When the United States entered World War I in April, 1917, the Army was unprepared for gas warfare. It possessed neither supplies of gas masks nor standard chemical warfare agents, yet two years had elapsed since the German army had first used chlorine gas at Ypres. In addition, from 1915 to 1917 the governments of Germany, France, and Great Britain had enlisted the services of chemists in their respective countries in establishing programs of chemical warfare and had learned that research was essential if an army was to gain an advantage on the battlefield by releasing a new, unexpected chemical agent.

American military men were skeptical about the value of chemical warfare in 1917. Their attitudes toward the intrusion of chemists and gas into warfare have been examined elsewhere (11). Other reasons for this lack of preparation are that improved gas masks had reduced the effectiveness of poison gas as a weapon, and at the same time extensive cloud gas attacks like the first attack at Ypres had become increasingly difficult to carry out.

American officers only recognized the importance of chemical warfare later in 1917, when two major innovations were introduced at the front. It was then that the German army first used mustard gas (dichlorodiethyl sulfide), a skin vesicant that produced a very high rate of casualties since masks were of little protection, and the British army first employed the Livens projector, a mortar capable of hurling a 150-pound drum of toxic chemicals into the enemy's trenches (12). So it was not before late summer of 1917 that the United States War Department considered seriously the development of a program for chemical warfare research. By then such a program already was under way, conducted by a few men associated with the Bureau of Mines who took the responsibility of assigning specific research projects to chemists in the colleges and universities of the country.

The National Research Council had been created in 1916 in anticipation of the war specifically to serve as an intermediary to provide the government's Council of National Defense with technical advice from the scientists of the country. However, few of the NRC committees had worked out plans for this function when the United States entered the war. A proposed survey of educational institutions to make an inventory of facilities, equipment, and manpower had not yet begun at the time when many university and college faculty were offering their services to the government (13, 14). Throughout 1917 the American Chemical Society urged academic chemists not to enlist in the army but wait until the government could accept their services as chemists (15). Meanwhile, the Bureau of Mines had already assumed responsibility for conducting research on chemical warfare and had enlisted the assistance of many chemists throughout the country. The Bureau's program was well established when it was placed for administrative purposes under the NRC's committee on noxious gases.

Since its establishment in 1910 as a bureau of the Department of the Interior, the Bureau of Mines had maintained on its staff a number of scientists and engineers who studied the toxic and asphyxiating gases often found in mines. Considering this expertise to be of value in the area of gas warfare, the director of the Bureau of Mines, Van H. Manning, offered the services of his bureau to the military committee of the NRC on February 8, 1917 (16). Although his offer was not accepted for several months, Manning proceeded with plans to begin research on chemical warfare in the Bureau of Mines. The Bureau's small research facility, the Pittsburgh Experiment Station, could not accomodate a large research program, and so Manning decided to seek volunteers among chemists of the United States who would conduct investigations in their own laboratories. Accordingly, he initiated an inventory of available chemists and engineers in cooperation with the American Institute of Mining Engineers and the American Chemical Society. Their members received questionnaires on which they were to report their areas of expertise, their willingness to participate in war-related projects, and the amount of time they could contribute (17, 18). By July, 1917, information on over 15,000 persons was available for the selection of those volunteers who would be asked to conduct specific research projects. Manning considered the Bureau of Mines file to be a nearly complete list of the chemists of the country since the membership of the American Chemical Society at that time was 14,500. During the course of World War I Manning's file often was consulted by other governmental agencies that were recruiting scientists, and most of the chemists who later entered military service were identified through this file and transferred to units where they could serve as chemists.

The military committee of the NRC finally accepted Manning's offer on April 3 and created a committee on noxious gases with Manning as chairman (19). It was through this committee of the NRC that the Bureau of Mines received its authority to conduct research on chemical warfare. Manning, as chairman, conducted its function nearly entirely through the Bureau of Mines organization until the NRC committee on noxious gases was dissolved on August 10, 1918. In selecting an administrative staff Manning primarily drew upon Bureau of Mines staff and a few industrial and academic chemists who had served as consultants to the Bureau in the past.

At the second meeting of the committee on noxious gases, held on April 21, Manning centered the discussion on the best plan for the organization of the research program. He explained to the 18 members present (his staff, several Army and Navy officers, and members of the chemistry committee of the NRC) that in connection with his inventory of chemists, offers were coming in from many individuals and laboratories throughout the country. For example, chemists at the Johns Hopkins University had offered the free use of their services and the university's laboratories. Manning thought they should take advantage of these offers rather than conduct the chemical warfare research exclusively in Bureau of Mines laboratories. The discussion then moved to the way research on chemical warfare was organized in Great Britain, the disadvantages of the British division into offensive and defensive committees, and the duplications of effort which had resulted from excessive secrecy and compartmentalization of research. The decision was reached that one central committee would be kept current on the progress of all research and that several subcommittees would be created representing specific projects, for example one to investigate new offensive agents. The subcommittees were to assist the central committee in defining research problems and assigning them to volunteers. All reports were to be sent to the central committee who would ensure that duplication did not occur and that all of the specific projects assigned were proceeding together.

George A. Burrell, a former director of the Bureau of Mines Pittsburgh laboratory, was selected by Manning to be in charge of research, and he was given the immediate task of compiling a list of specific research problems for distribution to volunteers. Since the NRC had made it clear to Manning that he was to interfere as little as possible with chemical industries, he turned his attention to chemists in colleges and universities (20). Throughout the month of May, two of Manning's staff, Warren K. Lewis and Bradley Dewey, traveled to institutions in the East and Midwest, showing to chemists the long list of problems compiled by Burrell and enlisting their aid. On May 25, 1917, the Bureau of Mines accepted the assistance of 122 chemists and established 28 branch laboratories of the Bureau of Mines (20 in colleges and universities, four in industrial laboratories, and four in governmental laboratories).

These laboratories and the number of chemists working in each were as follows: Carnegie Technical School, 5; Massachusetts Institute of Technology, 2; The Johns Hopkins University, 6; Harvard University, 5; University of Chicago, 4; Ohio State University, 5; University of Wisconsin, 4; University of Illinois, 4; University of Cincinnati, 1; New York University, 4; City College of New York, 2; Columbia University, 3; Massachusetts Agricultural College, 2; Amherst College, 1; Worcester Polytechnic Institute, 4; University of Pittsburgh, 2; Cornell University, 4; New Hampshire University, 1; Princeton University, 2; Bryn Mawr College, 1; Bureau of Mines, 4; Bureau of Chemistry, 4; Bureau of Standards, 2; Forest Products Laboratory, 6; Mellon Institute, 4; Pittsburgh Testing Laboratory, 2; National Electric Lamp Co., 3; National Carbon Co., 7 (21). These laboratories constituted the initial sites for project-research on chemical warfare. In order to carry out their plan to have one central research committee that would coordinate the specific problems assigned to these laboratories, the committee on noxious gases drew up plans to construct a central laboratory in Washington, D.C.

The branch laboratory at the Johns Hopkins University became the most important one, since in the course of its work many additional chemists of the country were recruited. Of the four members of the chemistry department who volunteered for projects, E. Emmet Reid was the most active in this regard. His assignment was to search among organic compounds for those that might be suitable new chemical warfare agents. He was authorized by the Bureau of Mines to employ six graduate students as assistants, their salaries being paid by the Bureau (22). Reid prepared a list of substances which appeared promising and set out to synthesize them and evaluate their chemical and physical properties. This task proved to be too large for his laboratory, and during the summer of 1917 Reid distributed the task of preparation of specific compounds to over 60 chemists throughout the country and maintained an active correspondence with them. Many of these men had been his students or had received their doctoral degrees from the Johns Hopkins University. Reid asked them to suggest additional compounds to him and assigned the synthesis of these compounds and those with related structures to other volunteers. In some cases the duplication of effort was deliberate; Reid gave the same synthetic problem simultaneously to several chemists. They submitted reports and in many instances mailed samples of the highly toxic compounds to Reid. Fifty-six such samples were received by him during the course of the war. Reid's laboratory prepared an additional 70 compounds and submitted samples of all of these for toxicological testing as they were prepared. Based on the reports of these tests Reid attempted to correlate structure and toxicity and as a result suggested many related compounds which he added to the original list of compounds to be synthesized. Reid was appointed a consultant to the Bureau of Mines, initially at a salary of one dollar a year. He preferred not to accept a commission in the Army, yet such commissions increasingly became common among chemists in the branch laboratories as a way of receiving governmental payment for their services. For example, at the end of the war there were 34 soldier-chemists stationed at the Johns Hopkins University laboratory (23).

The branch laboratory established at Yale University was given responsibility for testing the toxicity of samples prepared by the group of chemists organized by Reid. Throughout 1917, Yandell Henderson, a Yale physiologist and a consultant to the Bureau of Mines, directed the activities of the staff of about 50 civilian and military personnel at this station (24).

The branch laboratory at the University of Wisconsin was assigned the study of the effects of prolonged exposure to low concentrations of some of the more important toxic compounds, with the object of determining the hazards which would accompany their manufacture on a large scale. At the height of its activity this laboratory involved 15 members of the faculty. Some were commissioned in the Army, but most served as civilian consultants to the Bureau of Mines at a token salary. They were assisted by 33 soldiers who were assigned to duty at the University of Wisconsin. Later in the war, the work was expanded to include the study of the medical effects of gas poisoning and the development of salves for protection against mustard gas (25).

The largest of the branch laboratories was established at Catholic University in Washington, D.C. Its staff of about 75 carried out research on arsenic compounds and subsequently developed one of the new war gases produced by the United States during World War I, Lewisite (26).

The moisture and gas content of charcoals and the activation of charcoal for use in gas masks was the major project undertaken by the branch laboratory at Princeton University. George A. Hullet, a professor of physical chemistry at the University, directed a staff of 14 chemists who as soldiers were stationed there during the war. Fred Neher, an organic chemist, was assisted by three graduate students employed by the Bureau of Mines in the synthesis of several compounds suggested by E. Emmet Reid (27).

The work of many of these laboratories was coordinated during the summer of 1917 by James F. Norris, a chemist of the Massachusetts Institute of Technology, who visited each regularly to provide information on the progress of the other laboratories and to suggest additional projects. The regular reports of Reid and Henderson pointed out several compounds the chemical and physiological properties of which indicated that they would be useful as war gases, but Norris found it difficult to assign to the branch laboratories the much more dangerous tasks of setting up small-scale synthetic plants for the production of these agents. For this reason and because of the large amount of time spent by Norris in visiting branch laboratories, Manning pressed on the construction of the central laboratory he had envisioned from the beginning.

In September of 1917 the Bureau of Mines experiment station at American University in Washington, D.C. began operations. American University gave the use of its buildings to the Bureau for the duration of the war and this site grew to become the main center for chemical warfare research in the United States during World War I. Within a few months a large number of chemists in the Army were assigned to this station, having been identified through the Bureau of Mines file of chemists. In addition, many academic chemists were being released from their teaching duties in order to serve as full-time consultants to the Bureau of Mines, and they came to this experiment station to contribute to the expanding work on chemical warfare.

Project-Research on Chemical Warfare

On September 21, 1917, the United States Army stated that in view of the growing number of soldiers being assigned to war gas research, the work could be administered better under military control (28). Manning and other leaders of the Bureau of Mines who had created the research program opposed this transfer to military control and cited the accomplishments of the efficient organization they had created (16). The American Chemical Society also defended the civilian organization under the Bureau of Mines and expressed fear that control by the military might suppress "the spirit of originality, daring and speed in following new trails, so essential to the successful prosecution of research" (29). Despite these protests, chemical warfare research in the United States was transferred to the Army on June 26, 1918. This transfer actually did not alter the policies or the administration of research, but it did cause most of the chemists who were designated consultants to the Bureau of Mines at a token salary to accept commissions in the Chemical Warfare Service, a newly created branch of the Army. The Chemical Warfare Service brought all elements of chemical warfare under one administrative head, from the chemists serving in various gas regiments operating in combat in France to those in the Research Division designing new chemical agents. With this transfer the Bureau of Mines staff of 1682 scientists at the American University Experiment Station and at the branch laboratories throughout the country became the Research Division of the Chemical Warfare Service. Of this number, 1,034 were designated as technical personnel, mostly chemists, and 648 as non-technical, auxiliary personnel (16, 30). On November 11, 1918, the number of chemists in the Armed Forces numbered 5,404 and 4,003 of these were employed in chemical work (31). A large proportion of the chemists (about 2000 of these) were members of the Research Division and most of these were engaged in the type of research that had become characteristic of the research on chemical warfare, project-research.

The Research Division of the Chemical Warfare Service with headquarters at American University conducted research on all aspects of chemical warfare; including the study of various charcoals and other absorbents for gas, initial selection and synthesis of possible chemical agents, testing these for toxicity, and development of small-scale manufacturing processes. This program was divided initially into ten general research sections, including offense, pharmacological research, defense, editorial work, and small-scale manufacturing. Individual projects were assigned to small groups of investigators at the central laboratory and branch laboratories by the section chiefs, most of whom were at the central laboratory during 1918. The section chiefs prepared reports every two weeks and summary reports less frequently for the chief of the editorial section, Wilder D. Bancroft. Bancroft's staff prepared summaries of the relative progress on the various projects and pointed out the significance to other research of developments in specific projects. These summaries were circulated to the various research sections as well as other divisions of the Chemical Warfare Service. The editorial section also prepared replies to reports received from the British and French laboratories and prepared summary reports of American progress on problems of common interest to the Allies, such as a method for the manufacture of mustard gas.

The work of the offense section of the Research Division is illustrative of the method of project-research and it also shows the cooperation established between chemists in this section and the physiologists and pharmacologists in the pharmacological research section; this relationship was cited after the war as an example of cooperative research that might be useful in promoting research on new drugs. The work of the offense section actually began in May, 1917, when E. Emmet Reid was asked to undertake the investigation of organic compounds and to prepare such ones as were suitable for chemical warfare. As stated previously, Reid assigned the synthesis of specific compounds or classes of related compounds to particular individual chemists and to groups in the branch laboratories as well as chemists working in his laboratory at the Johns Hopkins University. After the work was transfered to control of the Army the work of the new offense section was primarily centered at American University, but projects continued to be assigned to the volunteers Reid had organized. In June, 1918, the offense section consisted of a staff of 165 at the central laboratory at American University, but also 50 chemists in seven of the branch laboratories and also a few scattered individual chemists in other universities. The branch laboratories associated with this section were at Worchester Polytechnic Institute, Bryn Mawr College, the Johns Hopkins University, Princeton University, Ohio State University, Yale University, and Columbia University (23, 32). In June, 1918 Lauder W. Jones, formerly of the University of Cincinnati, was appointed chief of the section.

A general pattern emerged in the offense section whereby whenever a new research problem arose it would be given to a team of investigators at one of the above laboratories. The central laboratory at American University had six research units, three of these were devoted to organic chemistry, two to analytical chemistry and one to inorganic chemistry. Each of the seven branch laboratories above constituted an additional unit. For example, the task of devising a method for the synthesis of mustard gas was assigned to a unit of organic chemists at American University under the direction of James B. Conant (33). Conant divided this problem into the study of three synthetic processes: 1) the production of ethylene; 2) ethylene chlorohydrin from ethylene; and 3) mustard gas from ethylene chlorohydrin. For the latter project Conant enlisted the aid of Moses Gomberg at the University of Michigan, and Gomberg subsequently submitted from his laboratory at Michigan a report on the synthetic process he developed (34). An alternate method for the synthesis of mustard gas was achieved by Conant's unit in cooperation with a British research team. It was typical of project-research in the Research Division that the same task was given to different groups so that they might pursue independent paths and results would be attained in the shortest time.

Cooperation between teams of investigators was important on projects involving the search for new chemical warfare agents. During a period of less than two years the chemists of the offense section prepared over 1600 compounds for examination as possible war gases (35). This was made possible through the close association of chemists with members of the pharmacological research section throughout the initial stages in the development of a promising new chemical agent. The general procedure was as follows. A sample of the new compound was prepared by one team, then its physical and chemical properties were studied by another team to determine its stability in shell casings, its volatility, etc. Meanwhile, a unit of the pharmacological section examined its biological effects. A meeting was then arranged between the chemists, physiologists, and pharmacologists involved, to decide if the compound was suitable for use in combat. If so, the task of developing a small-scale manufacturing process was given to another section, and at the same time a sample of the compound was examined by the defense section to ensure that American gas masks could be modified to give adequate protection against this compound. If all investigations to this point gave satisfactory results, the project was submitted to the Development Division or directly to the Gas Manufacturing Division of the Chemical Warfare Service. Often the chemists who were most familiar with the new agent were transferred to these divisions with the project. The efficiency with which this arrangement was conducted can be illustrated by the short time required for the scientists of the offense section to select, prepare, study, and finally produce on a small scale, the three chemical agents which were original contributions of America's chemical warfare research.

Lewisite was the best known of the war gases developed by the Research Division. Although it was not used on the battlefield, it became the subject of many sensational reports after the war. Lewisite was named for Winford Lee Lewis, a chemist at Northwestern University, who first isolated it and studied its properties as a vesicant. In January, 1918, an additional branch laboratory of the offense section was established at Catholic University, and it was here that Lewis did his work as leader of "organic unit no. 3." There was great interest in arsenic compounds following the introduction of chlorodiphenylarsine, "sneeze gas," by the German army. In a survey of the chemical literature, Lewis read of an arsenical produced from the reaction of arsenic trichloride and acetylene. He found subsequently that one of the several products of this reaction had excellent properties for use as a war gas. During the summer of 1918, Lewis's unit and Conant's unit sought methods for the efficient synthesis of this new agent. Four months after this project began, a plant for the manufacture of Lewisite according to a method worked out by Conant and Lewis was under construction.

In order to provide an example of project-research it is useful to examine in more detail the coordination of the work of the two units of the offense section and other parts of the Research Division on the various projects associated with the development of Lewisite. On April 13, 1918 Lewis read in an unpublished dissertation completed in 1904 at Catholic University by Julius Arthur Nieuwland of a poisonous, gummy mass produced in the reaction of acetylene and arsenic trichloride in the presence of aluminum trichloride. Six of the 12 to 14 members of organic unit no. 3 at this time were assigned by Lewis to the study of this reaction. Within two weeks they established that the gummy mass consisted of three compounds. Samples were isolated of each, they were purified, and sent to the pharmacological section for testing. Meanwhile the chemists determined the chemical identity of the three substances. It was reported by the pharmacological section that all three were vesicants and poisons but not to an equal degree. Because of the potentcy of one of the three, dichloro (2-chlorovinyl) arsine, James F. Norris, then head of the offense section, increased the size of organic unit no. 3 to 35 by the assignment of additional soldier-chemists to the laboratory at Catholic University. The entire team concentrated their efforts on increasing the proportion of what came to be called Lewisite that was produced in the reaction, preparing various derivatives of this compound, and substituting related acetylenes in the reaction. Soon, Norris called upon Conant's unit at American University to scale up the synthetic process. Finally this task was transferred to the Development Division which constructed a manufacturing plant at Cleveland. The first shipment of Lewisite was enroute to Europe when the Armistice was signed, and the dangerous cargo was dumped at sea (26, 32, 36).

Adamsite, 10-chloro-5, 10-dihydrophenarsazine, was named for Roger Adams, a chemist at the University of Illinois, who during the war was in charge of organic unit no. 2 at American University. On March 1, 1918, this unit was formed and given the project of developing an efficient synthesis for chlorodiphenylarsine, since at that time the method used by the Germans was unknown to the Allies. At a conference on April 18, David Worral, formerly a chemist at Smith College, suggested the preparation of a compound with a similar structure, 10-chloro-5,10-dihydrophenarsazine, the synthesis of which seemed much simpler (32). The chemists under Adams prepared this substance, and it was found that it also was an effective "sneeze gas" when dispersed as a smoke from flares (37). This smoke could penetrate gas masks, force the wearer to remove the mask because of the violent sneezing it caused, and thus expose the soldier to a lethal gas present on the battlefield. As in the case of Lewisite however, Adamsite was not produced in time to be used during World War I.

Chloroacetophenone was among the many samples of possible war gases prepared by E. Emmet Reid and sent to the Bureau of Mines in 1917. Because there were no testing facilities for lachrymators until the central laboratory was completed, the value of this compound as a tear gas went unnoticed. It was January, 1918, before the results of the physiological tests were reported which showed chloroacetophenone to be superior to any other tear gas in use at the time (23). The Johns Hopkins University branch laboratory, in cooperation with a unit at American University then developed a method of synthesis. Although chloroacetophenone was not produced in quantity before the war ended, it became the standard tear gas used by civilian police after the war (38).

Transfer of Project-Research to Peacetime

The experience of participation in project-research was shared by nearly 2000 members of the Research Division of the Chemical Warfare Service. Some of the most influential chemists of the country, in addition to pharmacologists and physiologists, were among this group, and after the war they rapidly returned to their former positions in government, industry, and higher education. They had learned the value of project-research in promoting the collaboration of scientists in different fields and working in laboratories separated by some distance. Those who previously specialized in organic, physical, biological, or analytical chemistry were united in the Research Division in their work on aspects of a single problem. Consequently, each appreciated more fully the advantages of working closely with their colleagues. From the reports issued by the editorial section many of the chemists came to appreciate other areas of the research in which they were not directly involved, and they were enabled to appreciate the intense research effort that was made on chemical warfare by scientists in the United States. As the accomplishments and methods of their wartime work became known through their personal contacts and published papers, the Research Division became known to an even larger number of scientists as a model of project-research.

The American Chemical Society took pride in the role it had played in the recruitment of chemists for research on chemical warfare and it was largely responsible for the publication of the results of their work. A series of articles appeared in the widely read Journal of Industrial and Engineering Chemistry designated as "Contributions from the Chemical Warfare Service," summarizing the techniques and findings useful to the wider study of chemistry (39). When the War Department attempted to abolish the Chemical Warfare Service in 1919, the ACS cooperated in a campaign of publicity about the work of the Chemical Warfare Service and contributed in a major way to its survival (40). Many chemists who formerly had worked in the Research Division delivered public addresses and wrote letters in support of the continuance of the Chemical Warfare Service to newspapers and to members of Congress.

The pattern of project-research was thus impressed on the consciousness of the chemical profession; it appeared to be capable of increasing the productivity and efficiency of science. The persistence of the pattern was evident in the policies established by the National Research Council, especially in its division of chemistry and chemical technology. In 1919 the coordination of the efforts of chemists in colleges and universities throughout the country on problems selected by the chairman of this division was regarded as a similar task to that dealt with in wartime by project-research. Even before the war ended, the type of research inaugurated in Manning's committee on noxious gases began to be used in other programs of the National Research Council. John Johnston was selected to become chairman of the division of chemistry and chemical technology at the beginning of 1918. Prior to this, as a member of Manning's committee he had been in charge of the offense section at American University (28). In his new post as head of the chemistry division Johnston employed the same pattern of project-research in order to effect collaboration of the nation's chemists with governmental laboratories. In 1918 he assigned about 50 specific projects to chemists in industry and educational institutions concerning the supply of essential chemicals for the war (41). In 1919, as the divisions of the National Research Council were reorganized on a permanent basis, Johnston was succeeded as chairman by Wilder D. Bancroft, a physical chemist at Cornell University. During the war Bancroft had headed the editorial section of the Research Division of the Chemical Warfare Service, and had coordinated the work of various research teams through his progress reports. Bancroft knew more than anyone about the pattern of project-research on chemical warfare, and in 1919 he produced the most comprehensive report of the entire operation (28).

Under the leadership of Johnston and Bancroft during the first two years of its operation in peacetime, the policy of the division of chemistry and chemical technology of the National Research Council became: "(1) to formulate special research problems for individuals; (2) to formulate general problems of fundamental importance involving cooperative work; (3) to bring together men in different branches of chemistry or men in different branches of science, each of whom is familiar with a given subject from a point of view not held by the others" (42). At the division's annual meeting, held in March 1919, a resolution was passed "that it shall be the policy of the division to form committees only for the purpose of undertaking definite projects" (43). Each committee was to assemble a list of important projects which could be undertaken in university laboratories, to ascertain for each project an institution which was able and willing to conduct the investigation, and to seek funds from chemical industries to support the investigations. At their next meeting the executive committee approved a questionnaire to be sent to all chemists of the United States on which they could indicate their desire to be informed of projects of the National Research Council within a specific field of chemistry (44).

It was Bancroft's opinion that to justify its existence, the division must produce results which could not be obtained otherwise, and that the best way to accomplish this was to select projects on the "borderlands" between chemistry and one or more of the other sciences (45). In accordance with this policy the majority of committees organized in the 1920s concerned these borderlands. One of the first to be formed was the committee on the chemistry of colloids, and in 1921 Bancroft published a list of 200 research projects in colloid chemistry for which volunteers were sought by the National Research Council (46). This committee published numerous monographs and arranged symposia on colloid chemistry, and was therefore influential in the establishment of an area of research within physical chemistry with large industrial importance.

In the selection of projects for 1923 emphasis was placed on obtaining suggestions from chemical companies for projects which were of fundamental scientific importance and would therefore be of interest to academic chemists. As chairman of the division of chemistry and chemical technology in 1923, J. E. Zanetti consulted industrial chemists and published a list of 100 selected research projects (47). As chairman, Zanetti maintained a correspondence file and a master list of problems on which he noted who had accepted the assignment of a problem (48). Zanetti also had been familiar with project-research in the Chemical Warfare Service, for as a liaison officer during the war he was responsible for the coordination of American and French research on chemical warfare.

This procedure of obtaining problems for investigation from industry and matching them to volunteers in colleges and universities was continued by the next chairman, James F. Norris, who also had served in the Research Division of the Chemical Warfare Service. In his final report as chairman, issued April 15, 1925, Norris described a new list of problems that was being compiled as a result of his recent visits to industrial research laboratories (49). The next chairman, William J. Hale, the director of organic chemical research at Dow Chemical Company, found the system of assigning problems to academic chemists unworkable due to their lack of familiarity with the methods of industrial research, and he dropped this program in 1925 (50). Within several committees of the division the method of project-research continued. In particular this method was used with great success for the compilation of the International Critical Tables of Physical and Chemical Constants (51). The use of project-research can be attributed primarily to the presence of a large number of former chemical warfare researchers who were active in the division of chemistry and chemical technology. During the period from 1918 to 1929, of the nine men who served as chairmen, six had participated in research on chemical warfare during the war (John Johnston, Wilder D. Bancroft, J. E. Zanetti, James F. Norris, William J. Hale and George Hulett), as had 20 of the 64 members of the division during this period.

The way in which the pattern of project-research was carried forward by the committees of the division can be illustrated by the work of one of the largest of these, the committee on chemical research on medicinal substances. This committee was established in 1923 and was continued until 1926 with Marston T. Bogert as chairman. As one of the original members of Manning's committee on noxious gases, it was Bogert who in 1917 suggested the establishment of branch laboratories to carry out project-research at colleges and universities. He subsequently served as chief of the relations section of the Chemical Warfare Service, and in that role he facilitated cooperation between the Chemical Warfare Service and colleges and universities (28). The committee on chemical research on medicinal substances was composed of 14 members who coordinated the work of an additional 57 collaborators attached to 12 subcommittees according to the type of compounds studied. Five of these 14 members (Marston T. Bogert, Roger Adams, Treat B. Johnson, W. Lee Lewis, and R. R. Renshaw) had participated in research on chemical warfare during the war. The extensive work of this committee resulted in the publication of 144 separate papers on the synthesis and testing of compounds for possible medicinal value (52).

The subcommittee on organic arsenicals included W. Lee Lewis, who had developed Lewisite and was then an organic chemist at Northwestern University, and Arthur S. Loevenhart, a pharmacologist at the University of Wisconsin. Loevenhart had been chief of pharmacological research at the central laboratory of the Chemical Warfare Service at American University. Loevenhart had been impressed by the cooperation established between organic chemists and pharmacologists in the search for new agents of chemical warfare, and after the war he proposed to continue this cooperation in a search for new drugs (53). In 1918 Loevenhart and Lewis collaborated on a project to synthesize and test the medicinal value of organic arsenicals. Their work was supported initially by the United States Interdepartmental Social Hygiene Board and later by the Public Health Institute of Chicago (54). Their project was one of many similar studies which were incorporated into the work of the subcommittee on organic arsenicals in 1923. A report issued by the subcommittee in 1925 contained a list of research projects and an invitation for chemists and pharmacologists to volunteer to study them (55).

The idea that project-research was well suited to the study of drugs also was present in the initial proposal in 1918 for the establishment of the National Institute of Health. Charles Holmes Herty's editorial in the *Journal of Industrial and Engineering Chemistry* entitled "War Chemistry and the Alleviation of Suffering," pointed out that the methods involved in such studies were identical with those used previously by the Research Division of the Chemical Warfare Service (56). (For Herty's role in establishing an institute for drug research see John Parascandola's chapter in this volume.) A committee of the American Chemical Society prepared a statement in 1921 of their plans and policies for a chemical-medical institute. Four of the nine members of the committee (Raymond F. Bacon, Reid Hunt, Treat B. Johnson, and Julius Stieglitz) had been associated with the Research Division of the Chemical Warfare Service. Their report commented on the pattern of scientific research employed in the research on war gases and recommended a similar organization for the battle against disease. The report concluded with the statement:

> Consequently it is proposed that the attack be actually cooperative, from the selection of the problem and the formulation of the plan of work through the whole concentrated effort to grapple with Nature and ultimately to conquer outpost after outpost of the complex world of life. May the day come when the lesson of the power of cooperative scientific endeavor, so effectively utilized in the Chemical Warfare Service organization, may be applied with equal success to the solution of the problems of disease and health (58).

Conclusion

Project-research proved to be an effective method of organizing research on chemical warfare during World War I. It solved a problem encountered early in the United States preparation for chemical warfare, that of efficiently coordinating the work of volunteer chemists who were isolated in separate laboratories. Later in the war it was continued more because it enabled the examination of the chemical, pharmacological, and technical aspects of one problem to proceed simultaneously in several research units. In this way results were expected to be obtained in the shortest possible time.

It is suggested here that this experience left its imprint not only on the chemists and other scientists involved, but also on the institution that they shaped to organize chemical research after the war. When the leaders of the division of chemistry and chemical technology of the National Research Council faced identical problems in the 1920s, they may well have turned to solutions they had found to be successful during World War I, yet evidence given here for the influence of the model of project-research is largely circumstantial. Many of the leaders of American Chemistry in the 1920s shared the common experience of research on chemical warfare, and their methods of organizing voluntary work on research projects in chemistry mirrored the wartime organization of the Chemical Warfare Service. Only biographical studies of these individual chemists can illuminate the direct influence of the model of project-research on their later careers. In this way the impact of the model on postwar chemistry might be better established, and the significance of research on chemical warfare for the development of chemistry in the United States might be demonstrated conclusively.

Literature Cited

1. Yerkes, Robert M., ed.; "The New World of Science: Its Development During the War"; Century: New York, 1920; Chapters 2, 3, 24.
2. Millikan, Robert A. *Science* 1919, *50*, 285-297.
3. Kevles, Daniel J. *ISIS* 1968, *59*, 427-437.
4. *Proc. Natl. Acad. Sci. USA* 1919, *5*, 188.
5. Jones, Daniel P., Ph.D. Thesis, University of Wisconsin, Madison, Wisconsin, 1969.
6. Baker, Newton D. *J. Ind. Eng. Chem.* 1919, *11*, 921.
7. Jones, Daniel P. *ISIS* 1980, *71*, 426-440.
8. Dupree, A. Hunter. "Science in the Federal Government"; Harvard Univ.: Cambridge, Mass., 1957; pp. 158-163.
9. Armsby, Henry Prentiss. *Science* 1920, *51*, 37.

10. Kellogg, Vernon. Educ. Review 1921, 62, 366-367.
11. Jones, Daniel P. Proc. Colloquium on Military History, Balesi, Charles J., ed.; United States Commission on Military History; Chicago, 1979; pp. 50-60.
12. Fries, Amos A.; West, Clarence J. "Chemical Warfare"; McGraw Hill: New York, 1921.
13. Thompson, William O. Ohio State Univ. Monthly 1917 (April).
14. Kolbe, Parke Rexford. "The Colleges in War Time and After"; Appleton: New York, 1919; pp. xvii, 40.
15. Editorials, J. Ind. Eng. Chem. 1917, 9, 544-545, 1085.
16. Manning, Van H. "War Gas Investigations"; U.S. Bureau of Mines Bulletin No. 178-A; U.S. Govt. Printing Office: Washington, D.C., 1919; p. 2.
17. Fay, Albert H. "Preparedness Census of Mining Engineers, Metallurgists, and Chemists"; U.S. Bureau of Mines Technical Paper No. 179; U.S. Govt. Printing Office: Washington, D.C., 1917.
18. Parsons, Charles L. J. Ind. Eng. Chem. 1918, 10, 776.
19. The minutes of this and subsequent meetings of the committee on noxious gases are in the Records of the Bureau of Mines, Record Group 70, Entry 2, National Archives.
20. Lewis, Warren K. Chemical Warfare 1932, 18, 1118.
21. Report submitted at meeting of committee on noxious gases, May 25, 1917.
22. Reid, E. Emmet. Armed Forces Chemical Journal 1955, 9 (July-August), 38.
23. Reid, E. Emmet. "History of Offense Research, John Hopkins University Station", Historical Report No. H-149, Chemical Warfare Service, Edgewood Arsenal Technical Library.
24. Henderson, Yandell. "History of Research at Yale University", Historical Report No. H-150, Chemical Warfare Service, Edgewood Arsenal Technical Library.
25. "History of the University of Wisconsin Section, Medical Division, Chemical Warfare Service", Historical Report No. H-151, Chemical Warfare Service, Edgewood Arsenal Technical Library.
26. Lewis, Winford Lee. "Summary of Work Done in Organic Unit No. 3, Offense Research Section, CWS", Historical Report No. H-209, Chemical Warfare Service, Edgewood Arsenal Technical Library.
27. Hulett, George A. "History of the Research Division at Princeton University", Historical Report No. H-153, Chemical Warfare Service, Edgewood Arsenal Technical Library.
28. Bancroft, Wilder D. "History of the Chemical Warfare Service in the United States", Historical Report No. H-11, Chemical Warfare Service, Edgewood Arsenal Technical Library, p. 58.
29. Editorial, J. Ind. Eng. Chem. 1918, 10, 590.

30. "List of Bureau of Mines Personnel", Records of the Bureau of Mines, Record Group 70, entry 2, National Archives.
31. Bogert, Marston T.; Nichols, William H.; Noyes, A. A.; Stieglitz, Julius; and Parsons, Charles L. J. Ind. Eng. Chem. 1919, 11, 415.
32. Jones, Lauder W. "Offense Chemical Research Section, Summary of Achievements, 1917-1918", Chemical Warfare Monograph No. 55, Edgewood Arsenal Technical Library.
33. Conant, James B. "Progress Report, Organic Section, Feb. 18, 1918" enclosed in a letter from G. Burrell to W. McPherson, Feb. 26, 1918, Records of the Chemical Warfare Service, Record Group 175, National Archives.
34. Gomberg, Moses. "Report on the Preparation of Dichloroethylsulfide", Record Group 175, National Archives.
35. Burrell, George A. J. Ind. Eng. Chem. 1919, 11, 93-94.
36. West, Clarence J. "Pharmacological Data", Chemical Warfare Monograph No. 50, Chemical Warfare Service, Edgewood Arsenal Technical Library.
37. Manning, Van H. "Report, May 1918", Records of the Bureau of Mines, Record Group 70, National Archives.
38. Jones, Daniel P. Technology and Culture 1978, 19, 151-168.
39. J. Ind. Eng. Chem. 1919, 11, 5-12, 13-19, 93-110, 185-200, 281-296, 420-443, 513-540, 621-629, 721-723, 817-836, 1013-1019, 1105-1116; 1920, 12, 213-223, 1054-1069.
40. Editorials, J. Ind. Eng. Chem. 1919, 11, 506, 814-816h; 1920, 12, 2-3, 107, 314.
41. National Research Council, "Annual Report 1918"; U.S. Government Printing Office: Washington, D.C., 1919.
42. National Research Council, "Annual Report 1919"; U.S. Government Printing Office: Washington, D.C., 1920; p. 33.
43. J. Ind. Eng. Chem. 1919, 11, 481-483.
44. J. Ind. Eng. Chem. 1919, 11, 694-697.
45. Bancroft, Wilder D. J. Ind. Eng. Chem. 1920, 12, 911.
46. Bancroft, Wilder D. J. Ind. Eng. Chem. 1921, 13, 83-89, 153-158, 260-264, 346-351.
47. Zanetti, J. E. J. Ind. Eng. Chem. 1924, 16, 304-307.
48. These documents are in the archives of the National Academy of Sciences, Washington, D.C.
49. Norris, James F. "Report of Activities of the Division of Chemistry and Chemical Technology. February-April 1925"; Archives of the National Academy of Sciences, Washington, D.C.
50. Hale, William J. "Report of the Chairman of the Division of Chemistry and Chemical Technology, December 4, 1925"; Archives of the National Academy of Sciences, Washington, D.C.
51. Kraus, Charles A. Science 1933, 77, 552-554.
52. National Research Council, "A History of the NRC 1919-1933"; U.S. Government Printing Office: Washington, D.C., 1933.

53. Loevenhart, Arthur S. J. Ind. Eng. Chem. 1918, 10, 971-973.
54. Leake, Chauncey D. Science 1925, 62, 251-256.
55. Christiansen, Walter G. J. Ind. Eng. Chem. 1925, 17, 1270-1271.
56. J. Ind. Eng. Chem. 1918, 10, 673-674.
57. J. Ind. Eng. Chem. 1918, 10, 969-976.
58. American Chemical Society. "The Future Independence and Progress of American Medicine in the Age of Chemistry"; Chemical Foundation: New York, 1921, p. 80.

RECEIVED December 27, 1982

Appendix

Historical Publications of Aaron J. Ihde

This bibliography does not include book reviews, short pieces (mostly biographies) in encyclopedias, letters, notes, reports, newsletters, pieces on science education, publications based on experimental research, or independent publications of students arising out of work directed by Aaron J. Ihde.

Books

The Physical Universe: Readings and Exercises. College Printing and Typing Co., Madison, 1963, 88 pp.; 2nd edn., 1964, 133 pp.; 3rd edn., 1965, 153 pp.; 4th edn., 1968, 177 pp.

The Development of Modern Chemistry. Harper and Row, New York, 1964, xii + 851 pp. including illustrations, Appendixes, (Discovery of the Elements, Discovery of Natural Radioactive Isotopes, Radioactive Decay Series, Nobel Prize Winners in Chemistry, Physics, and Medicine), and Bibliographic Notes.

International Student Reprint. John Weatherill, Inc., Tokyo and Harper and Row, London, 1966.

Japanese Edition, Translated by Chikayosi Kamati, Kiychisa Fujii and Chie Fujita. Misuzu Shobo, Tokyo, vol. I (chs. 1-9), 1972; vol. II (chs. 10-17), 1973; vol. III (chs. 18-27), 1977.

Paperback Reprint. Dover Publications, New York, 1983.

Selected Readings in the History of Chemistry. (With Wm. F. Kieffer). Journal of Chemical Educ., Easton, PA, 1965, vi + 298 pp.

Joseph Priestley: Scientist, Theologian, and Metaphysician. (With Erwin N. Hiebert and Robert E. Schofield). Edited by Lester Kieft and Bennett R. Willeford, Jr., Bucknell University Press, Lewisburg, PA, 1980, 117 pp. Ihde contribution is titled "Priestley and Lavoisier," pp. 62-91.

0097-6156/83/0228-0187 $06.00/0

Journal Publications and Occasional Papers

Maple Sugar. A Bibliography of the Early Records II. (With H.A. Schuette). Trans. Wis. Acad. Sci., Arts, Letters, 38, 89-184 (1946).

The Inevitability of Scientific Discovery. Sci. Monthly, 67, 427-29 (1948).

Chemistry and the Spectrum before Bunsen and Kirchoff. (With Tillmon H. Pearson). J. Chem. Educ., 28, 267-71 (1951).

History of Chemistry at the University of Wisconsin. Isis, 42, 308 (1951).

The Early Days of Chemistry at the University of Wisconsin. (With H.A. Schuette). J. Chem. Educ., 29, 65-72 (1952).

The Duveen Library. (With Samuel A. Ives). Ibid., 29, 244-47 (1952).

Edmund Ruffin: Soil Chemist of the Old South. Ibid., 29, 407-14 (1952).

Beginnings of Chemical Education in Beloit, Lawrence and Ripon Colleges. (With Robert Siegfried). Trans. Wis. Acad. Sci., Arts, Letters, 42, 25-38 (1953).

Learning the Scientific Method Through the Historical Approach. School Science and Math, 53, 637-43 (1953).

Responsibility of the Scientist to Society. Sci. Monthly, 77, 244-49 (1953).

Faraday's Electrochemical Laws and the Determination of Equivalent Weights. (With Rosemary Gene Ehl). J. Chem. Educ., 31, 226-32 (1954).

Are There Rules for Writing History of Chemistry? Sci. Monthly, 81, 183-86 (1955).

Chemical Industry in Early Wisconsin. (With James W. Conners). Trans. Wis. Acad. Sciences, Art, Letters, 44, 5-20 (1955).

The Pillars of Modern Chemistry. J. Chem. Educ., 33, 107-10 (1956).

Antecedents to the Boyle Concept of the Element. Ibid., 33, 548-51 (1956)

The Development of Scientific Laboratories. *Science Teacher*, *23*, 325-27 (1956).

Chemical Industry, 1780-1900. *Cahiers d'histoire mondiale*, *4*, 957-84 (1958).

The Unraveling of Geometric Isomerism and Tautomerism. *J. Chem. Educ.*, *36*, 330-37 (1959).

Commentary on the Papers of Cyril Stanley Smith and Marie Boas. In Marshall Clagett, ed., *Critical Problems in the History of Science*, University of Wisconsin Press, Madison, 1959, pp. 519-24.

Biographical Sketches in *Dictionary of Wisconsin Biography*, State Historical Society of Wisconsin, Madison, 1960. William Willard Daniells, p. 94; John Eugene Davies, p. 97; George Calvin Humphrey, p. 182; Herman Ihde, p. 185; Franklin Hiram King, p. 207; Edward Kremers, p. 213; Stephen Pearl Lathrop, p. 223.

The Karlsruhe Congress: A Centennial Retrospect. *J. Chem. Educ.* *38*, 83-86 (1961).

Jöns Jacob Berzelius. Chapter 30 in *Great Chemists*, E. Farber, ed., Interscience, New York, 1961, pp. 385-402.

Michael Faraday. Chapter 35 in *ibid*., pp. 465-80.

American Chemists at the Century's Turn. S.M. Babcock, Harvey Wiley, Ira Remsen, T.W. Richards, and Edgar Fahs Smith. Chapter 58 in *ibid*., pp. 805-30.

History of Food Standards. *Wis. Dietitian*, *24*, no. 3, 5-6 (1962).

Are the Liberal Arts Worth Saving in a Scientific World? *Wis. Acad. Rev.*, *10*, 97-101, 139-43 (1963).

Paracelsus - Genius or Charlatan? *Sixth Annual Lecture Series in the Pharmaceutical Sciences*, *1962-63*, University of Texas, College of Pharmacy, Austin, 1963, pp. 4-12. Reprinted in *Texas J. Pharmacy*, *5* (1964). Reprinted in AJI, *The Physical Universe*, 2nd edn., 1964, pp. 33-38, and subsequent editions.

Serendipity (Quinine to 606). *Ibid*., pp. 13-23. Reprinted in *Texas J. Pharmacy*, *5* (1964). Reprinted in *Physical Universe*, 2nd edn., *et seq*.

Molds, Sulfa, and Science. *Ibid*., pp. 24-36. Reprinted in *Texas J. Pharmacy*, *5* (1964). Reprinted in *Physical Universe*, 2nd edn., *et. seq*.

Alchemy in Reverse: Robert Boyle on the Degradation of Gold. *Chymia*, *9*, 47-57 (1964).

La Chimie. Chapter XI in *Histoire Generale des Sciences*, R. Taton, ed., vol. 3, part 2, *Le XXe Siecle*, Presses Univ. de France, Paris, 1964, pp. 397-447. Also available in translation (see following item).

Chemistry. Chapter 10 in *History of Science*: volume 4, *Science in the Twentieth Century*, R. Taton, editor. Transl. by A.J. Pomerans, Basic Books, New York, 1966, pp. 251-85.

The Scientist and the Modern World. *Trans. Wis. Acad., Sci., Arts, Letters*, *53*, 1-7 (1964). Presidential Address to Wisconsin Academy.

The Basic Sciences in Wisconsin. *Ibid.*, *54A*, 33-41 (1965).

Isomers and Isomerization. Article in "The Encyclopedia of Chemistry," George L. Clark, ed., 2nd edn., Reinhold Publ. Co., N.Y., 1966, pp. 570-72; 3rd edn., Van Nostrand-Reinhold, N.Y., 1973, pp. 593-95.

The Development of Strain Theory. *Advances in Chemistry*, *61*, 140-62 (1966).

The History of Chemistry Program at the University of Wisconsin. *Badger Chemist*, no. 14, 14-16 (1967).

W. Schneider's View of the Border Areas between the History of Pharmacy and the History of Chemistry. A section in *Pharmaceutical Historiography*, Alex Berman, ed., Am. Inst. Hist. Pharmacy, Madison, 1967, pp. 33-38.

Pest and Disease Controls. Chapter in *Technology and Western Civilization*, vol. II, Melvin Kranzberg and Carroll W. Pursell, Jr., eds., Oxford University Press, New York, 1967, pp. 369-85.

Free Radicals and Moses Gomberg's Contribution. *Pure and Applied Chem.*, *15*, 1-13 (1967); also in *Free Radicals in Solution*, Butterworth's, London, 1967, pp. 1-13. Condensed version in *Chem. Engr. News*, *44*, no. 41, 90-92 (Oct. 3, 1966).

Chemical Analysis and the Growth of Biochemistry. *Actes du XI Congres International d'histoire des Sciences*, Warsaw-Cracow, 24-31, Aug., 1965, *4*, 143-47 (1968).

Theodore William Richards and the Atomic Weight Problem. *Science* *164*, 647-51 (1969).

An Inquiry Into the Origins of Hybrid Sciences: Astrophysics and Biochemistry. *J. Chem. Educ.*, *46*, 193-96 (1969). Dexter Award Address.

Foreward. *The Periodic System of the Chemical Elements, A History of the First Hundred Years*, by J.W. van Spronsen. Elsevier, Amsterdam, 1969, pp. ix-x.

History of the Pneumatic Trough. (With John Parascandola). *Isis*, *60*, 351-61 (1969).

Alder, Kurt. *Dictionary of Scientific Biography*, C.C. Gillespie, ed., Charles Scribner's Sons, New York, *1*, 105-6 (1970).

Babcock, Stephen Moulton. *Ibid*., *1*, 356-57 (1970).

J. Howard Mathews (1881-1970). *Wis. Acad. Rev.*, *17* (4),15-16 (1971).

Conflict of Concepts in Early Vitamin Studies. (With Stanley L. Becker). *J. Hist. Biol.*, *4*, 1-33 (1971).

Elvehjem, Conrad Arnold. *Dict. Sci. Biog.*, *4*, 357-59 (1971).

Euler-Chelpin, Hans Karl August Simon von. *Ibid*., *4*, 485-86 (1971).

The History of Nutrition in the late Nineteenth and Twentieth Centuries. *Proc. Conf. on Hist. of Biochem. and Molec. Biol.*, May 21-23, 1970, Boston, Amer. Acad. Arts & Sciences, pp. 150-4.

Stephen Moulton Babcock -- Benevolent Skeptic. *Perspectives in the History of Science and Technology*, Duane H.D. Roller, ed., U. of Oklahoma Press, Norman, 1971, pp. 271-82.

Let's Teach History of Chemistry to Chemists! *J. Chem. Educ.*, *48*, 686-87 (1971).

History of Chemistry at the University of Wisconsin. *Teaching the History of Chemistry, A Symposium*, Geo. B. Kauffman, ed., Akademiiei Kiadi, Budapest, 1971, pp. 171-78.

Funk, Casimir. *Dict. Sci. Biog.*, *5*, 208-9 (1972).

Gomberg, Moses. *Ibid*., *5*, 464-66 (1972).

Harden, Arthur. *Ibid*., *6*, 11-12 (1972).

Hart, Edwin Bret. *Ibid*., *6*, 135-36 (1972).

Baekeland, Leo Hendrik. Dictionary of American Biography, Supplement 3, 1941-45, Charles Scribner's Sons, New York, 1973, pp. 25-27.

Chittenden, Russell Henry. Ibid., pp. 162-64.

Schoenheimer, Rudolf. Ibid., pp. 693-94.

Gortner, Ross Aiken. Ibid., pp. 314-15.

Kahlenberg, Louis Albrecht. Dict. Sci. Biog., 7, 208-09 (1973).

Levene, Phoebus Aaron Theodor. Ibid., 8, 275-76 (1973).

Fischer, Emil. Biographic Encyclopedia of Scientists and Inventors, Edizioni Scientifiche e Techniche, Mondadori, Milan, 1974.

Funk, Kazmierz. Ibid.

Pettenkofer, Max Joseph. Ibid.

Analytical Chemistry and the Effectiveness of Food Laws. J. Chem. Educ., 51, 295-97 (1974).

Studies on the History of Rickets. I. Recognition of Rickets as a Deficiency Disease. Pharmacy in History, 16, 83-88 (1974).

Pauling, Linus. Encyclopedia Britannica, 15th edn., 1974, pp. 1094-95 in the Macropedia.

Early American Studies on Respiration Calorimetry. (With Jerry F. Janssen). Molecular and Cellular Biochemistry, 5, 11-16 (1974).

Adkins, Homer Burton. Dict. Am. Biog., Supplement 4, 1946-50, 1974, pp. 5-7.

Gomberg, Moses. Ibid., pp. 335-37.

Koch, Fred Conrad. Ibid., pp. 459-61.

Studies on the History of Rickets . II. The Roles of Cod Liver Oil and Light. Pharmacy in History, 17, 13-20 (1975).

Elementi. Enciclopedia della Chimica, Edicioni Scientifiche, 1975, Utet Sansoni Edizioni Scientifiche, Florence, pp. 579-84.

Atomic Weight Determinations. Dictionary of American History, Rev. Edn., Charles Scribner's Sons, New York, 1976, pp. 218-19.

Biochemistry, Ibid., pp. 302-4.

Fertilizers, Ibid., pp. 16-17.

European Tradition in Nineteenth Century American Chemistry. J. Chem. Educ., 53, 741-44 (1976).

Man's Historic Attitude Toward the Environment. Trans. Wis. Acad. Sci., Arts, Letters, 64, 223-33 (1976).

Adkins, Homer Burton. American Chemists and Chemical Engineers Wyndham D. Miles, ed., Am. Chem. Soc., Washington, 1976, pp. 5-7.

Babcock, Stephen Moulton. Ibid., pp. 14-15.

Burgess, Charles Frederick. Ibid., pp. 54-55.

Daniels, Farrington. Ibid., pp. 110-12.

Elvehjem, Conrad Arnold. Ibid., pp. 143-44.

Kahlenberg, Louis Albrecht. Ibid., pp. 259-60.

Mathews, Joseph Howard. Ibid., pp. 319-20.

Ruffin, Edmund. Ibid., p. 420.

Wiley, Harvey Washington. Dict. Sci. Biog., 14, 357-58 (1976).

Williams, Robert Runnels. Ibid., 14, 392-94 (1976).

Bancroft, Wilder Dwight. Dict. Am. Biog., Supplement 5, 1977, pp. 35-37.

Hudson, Claude Silbert. Ibid., pp. 327-28.

Edward Mellanby and the Antirachitic Factor. (With John Parascandola). Bull. Hist. Med., 51, 507-15 (1977).

Chemistry in the Old Northwest. Ohio J. Sci., 78, 59-69 (1978).

Louis Kahlenberg's Opposition to the Theory of Electrolytic Dissociation. Proc. Symposium on Selected Topics in the History of Electrochemistry, Geo. Dubpernell and J.H. Westbrook, eds. (Proc. vol. 78-6, 1978, The Electrochem. Society, Princeton, N.J.), pp. 299-312.

A Badger Chemist Genealogy: The Faculty at the University of Wisconsin. (With Alan J. Rocke). J. Chem. Educ., 56, 93-5 (1979).

Linus Pauling. Enciclopedia Universale Unedi - Dizionario Enciclopedico, Casa Editrice SCODE, Milan, 1979, vol. 10, p. 497.

Harry Steenbock - Student and Humanist. Wis. Acad. Rev., 26 (1), 15-17 (1979).

Kirkwood, John Gamble. Dict. Amer. Biog., Supplement 6, 1956-60, 1980, pp. 345-46.

Stern, Kurt Guenter, Ibid., pp. 595-96.

Chemistry is a Human Enterprise. J. Chem. Educ., 57, 11-12 (1980). Reprinted in Source Book for Chem. Teachers, W.T. Lippincott, ed., Div. Chem. Educ., Am. Chem. Soc., Washington, 1981, Ch. 5, "Historical Pieces and Anecdotes", A.T. Schwartz and Lois Fruen, eds., pp. 53-5.

Roberts, Lydia Jane. Notable American Women, The Modern Period, Barbara Sicherman, et al, eds., Belknap Press, Harvard U. Press, Cambridge, 1980, pp. 580-81.

The History of Acid-Base Theory. Audio Tape and Manual. ACS Audio Courses, Hist. of Chem. Series, Am. Chem. Soc., Washington, 1980, 42 pp., manual, 52 min. tape.

Elvehjem, Conrad A. Dict. Amer. Biog., Supplement 7, 1961-65, 1981, pp. 223-4.

Chemistry at the University of Wisconsin-Madison 1848-1980. Trans. Wis. Acad. Sci., Arts, Letters, 69, 135-52 (1982).

Dexter Award: Silver Anniversary. Isis, 73, 268-70 (1982).

Food Controls Under the 1906 Act. The Early Years of Federal Food and Drug Control, J.H. Young, ed., Am. Inst. Hist. Pharmacy, Madison, 1982, pp. 40-50

RECEIVED March 18, 1983

INDEX

INDEX

D

I

J

K

L

M

Indexing and production by Susan Robinson
Elements typeset by The Sheridan Press, Hanover, PA
Printed and bound by Maple Press Co., York, PA